GROWTH AND NUTRITIONAL STUDIES:
PERSPECTIVE FROM NORTHEAST INDIA

GROWTH AND NUTRITIONAL STUDIES:

PERSPECTIVE FROM NORTHEAST INDIA

Edited by

SARTHAK SENGUPTA
DALI DUTTA

GYAN

Published, 2019

Published by

Gyan Publishing House
5, Ansari Road
Daryaganj, New Delhi-110002
Phone: 011-47034999, 9811692060
www.gyanbooks.com
E-mail: books@gyanbooks.com

Printed at: G. Print Process, Delhi.

Cataloging in Publication Data—DK
Courtesy: D.K. Agencies (P) Ltd. <docinfo@dkagencies.com>

Growth and nutritional studies : perspective from Northeast India / [edited] by Dr. Sarthak Sengupta.
pages cm
Includes bibliographical references.
ISBN 9788121214933

1. Tribes—Nutrition—India, Northeastern. 2. Tribes—Health and hygiene—India, Northeastern. 3. Human growth—India, Northeastern. I. Sengupta, Sarthak, 1955- editor.

LCC GN296.5.I4G76 2018 | DDC 305.8009541 23

Contents

Prologue

The term growth refers to a change in size and is usually applied to the period from birth to physical maturity. The concept of physical growth is a multifaceted phenomenon. It is not just a steady and uniform process of increase in different parts of the body but exhibits sufficient variation in the increase during the preceding or the succeeding years. In the process, some portions of the body increase rapidly while others do not. Growth, therefore, is the study of change in an organism, which is not yet mature.

Nutrition is the direct outcome of food intake, food habits and is based on the economy, environment and the availability of particular food. Brown (1984) defined that nutritional status is a physical reflection of the interaction between individuals' nutritional intakes, the bio-availability of different nutrients, and his or her physiological requirements.

Pioneering work on growth study in Northeast India was initiated by Das (1966, 1971) studying the Kalita caste of Assam. The study recorded that Kalita girls are heavier, taller than the boys towards the early period, but at the later period the boys become heavier and taller than the girls. Later, she studied the growth of the Kaibarta boys (Das, 1973a), and Baishya children (Das 1973b) of Assam. She also carries out a comparative growth study among the boys of Brahmin, Kalita, Baishya, and Kaibarta (Das, 1974). It was observed that in all the castes adolescent growth spurts occur at the age of 12 to 13 years. Das and Das (1969-1971) have reported some aspects of the growth pattern of the Assamese boys belonging to the age group of 6 to 16 years. Ahmed Hazarika (1974, 1981) completed a growth study of the Ahom of Dibrugarh. He found that pre-adolescent growth spurt occurs at the age of 8 years, and adolescent growth spurt in between 13-14 years and post-adolescent growth spurt between 16-19 years of age. Das (1978, 1982) made a comparative study on the growth and body build of the Assamese and Rajasthani boys. It was noted that Assamese and Rajasthani boys are taller and heavier than the Indian urban boys throughout the growth period.

Choudhury (1979) studied the growth pattern of the Rabha boys of Assam. The study revealed that the Rabha boys were taller and heavier than the boys belonging to Assamese castes during the early age period. He also attempted to explain differences in growth patterns among the boys of Northeast India in respect of different nutritional status, environmental condition and genetic difference among the populations. Ashizawa and Kawabata (1984) studied the growth pattern and role of nutritional factors on the growth of the Muslim population from Kamrup District Assam. Choudhury and Das (1985) studied growth in height and weight of Assamese children. Choudhury et al. (1992) studied the growth pattern of school going children of Assamese, Bengali and Hindi speaking groups of Guwahati city. The Assamese children were found to be the tallest and heaviest. Choudhury and Bhuyan (1994) studied the growth and nutritional status of rural Kalita children of Kamrup district. The nutritional assessment showed that almost 60.0 percent of the Kalita children suffered from varying degrees of malnutrition. Das and Hazarika (1995) studied the height and weight of postgraduate students of Gauhati University.

Begum (1995a, 1995b) studied growth pattern and nutritional status of Assamese Muslim of Kamrup district. Choudhury and Gogoi (1995) studied the Neo-Assamese Muslim population of Kamrup district for their growth status in the pre-adolescent period. Begum and Choudhury (1997) did an assessment of nutritional status using both age-dependent and independent anthropometric parameters among Assamese Muslims of the same district. They also reported the socio-economic interactions with growth and nutritional status among the population. Both the researcher (Begum and Choudhury, 1999) also studied the growth in some somatometric characters of the Assamese Muslims of Kamrup district.

Dutta (1997) studied growth pattern and nutritional status in the early childhood of Kaibarta and Ahom populations of Dibrugarh. He found that Ahom children had a better nutritional status than Kaibarta children. Chelleng and Mahanta (1998) studied the growth and development of children of tea garden labourer of Assam. Das (2001) studied the growth and development of the Kaibarta children of Guwahati and North Guwahati. Kulkarni and Sarma (2001) reported nutritional status based on anthropometry among the Karbi tribe of Assam. Das and Boruah (2002) studied the growth and development of Ahom children in Sivasagar district. Sahani and Das (2003) carried out a similar study on the growth of adolescents of Ahom children in rural areas of Sivasagar district. Hussain (2003) studied the somatotypes, physical growth and nutritional status of the Bengali Muslim boys of the Barak valley, Assam. Khongsdier (2002) assessed the nutritional status of the Kaibartas of Assam.

Kotal and Boruah (2004) studied the physical growth of adolescents among the Sonowal Kacharis of Dibrugarh district. Medhi et al. (2006) assessed the growth and nutritional status of school-age children of tea garden workers of Assam. A high prevalence of malnutrition among tea garden children of Assam is reported. Sikdar and Ahmed Das (2006) studied growth progression with ages in physical and physiological variables among the Bengali population of Dibrugarh, Assam. Singh (2008) studied the growth and nutritional status of the Sonowal Kacharis of Dibrugarh district. Sikdar (2010) studied biosocial determinants of growth and nutritional status of the Mishings of Assam. Dutta (2012) studied the growth and nutritional status of Ahom children of upper Assam. Medhi (2013) studied on growth and nutritional status of Bengali speaking Muslim of Barpeta district and she also studies the impact of consanguinity on growth. Deori (2016) studied growth among the Deori populations of Assam and Arunachal Pradesh.

The first longitudinal study in entire Northeast India was done by Das (1977) dealing with the growth in height, weight and sitting height of five siblings for a period of ten years. The same researcher (Das, 1978) reported the findings of a longitudinal study on the growth of head and face dimensions of four sisters for twelve years. Later Das and Choudhury (1982, 1992) reported a mixed longitudinal study on height and weight of Assamese children and showed the variations in growth pattern among them. They found that growth in height was very fast from 1 to 3 years.

In Arunachal Pradesh, Duarah and Das (1978) made a growth study on Khamti boys. Duarah (1985) also studied the growth pattern in respect of some anthropometric characters of three groups of Monpa boys of Arunachal Pradesh inhabiting in varied altitudes. He observed that the distance curves for almost all the characters of the Dirang Monpa samples remain at higher levels all through than the other two samples. Mantaw and Priyadarshini (2000) did an overview of the nutritional status of children (1 to 5 years) in Arunachal Pradesh based on the data presented in India Nutrition Profile (1998). They reported the prevalence of underweight as 68.00 percent, severe undernutrition as 24.00 percent, stunting as 68.00 percent and 29.00 percent were wasted. The physical growth of Nishi adolescents of Papumpare district of Arunachal Pradesh was studied by Kotal and Boruah (2004).

In Manipur, Luwang (1981) studied the pattern of weight in children under 5 years of age of hill and valley areas of Manipur. Ashizawa and Kawabata (1984) carried out a study to assess the body size in relation to age at menarche of Meitei girls of Thoubal district. Devi (1986) studied the physical growth of the Meitei and the Muslims of Manipur. It is reported that the adolescent growth spurt occurs earlier in the Muslim than in the Meitei. Singh (1986) studied the Kabuis of hills and valley of Manipur in respect of certain anthropometric traits. The Kabui living in valleys show higher mean values in most of the traits. Gaur and Singh (1995) reported the pattern of growth among the rural school-going children with poor socio-economic backgrounds from Manipur. Talwar and Singh (1995) reported the growth pattern of adolescent Meitei children of Thoubal district, Manipur. The Meitei girls show more sub-cutaneous fat at all ages than boys do. Singh and Singh (2000) studied physical growth among the affluent and non-affluent Meitei boys of Manipur. It is reported that the affluent boys show higher mean values in almost all the measurements. Physical growth and nutritional status among school children of Thanga Island was studied by Singh and Konsam (2001). Jibonkumar and Singh (2006) reported the nutritional status of Sekmai children of Manipur. Haokip (2015) studied growth, body dimension and nutritional status among the Thadou Kukis of Chandel district, Manipur.

In Meghalaya, Das (1988) worked among the Hajongs. Their adolescent growth spurt was between 12 and 15 years of age. Khongsdier (1995, 1996) studied the growth pattern, nutrition and health status of the War Khasi children with regard to height for age and weight for height. It is reported that the adolescent growth spurt occurs at the age of 14 years in boys and 12 years in girls in respect of several measurements. Pathak (1995) evaluated age changes in stature and weight among adolescent Khasi girls. Mukherjee (2002) studied the growth pattern among the Christian, Muslim and *Niamtre* Khasi children of Shillong. The study reveals that the Khasi children are much shorter and lighter than the U.S. NCHS children but more or less comparable in terms of weight and height of children, data provided by ICMR. Khongsdier and Mukherjee (2003) studied the growth and nutritional status of urban Khasi boys of Meghalaya. About 60.00 percent, 29.00 percent, and 65.00 percent were below -2 Z scores of the US (NCHS) references in respect of height for age, weight for age and body mass index respectively. Muslim Khasi boys, who were the offspring of inter-marriages between Khasi females and immigrant Muslim males, were heavier and taller than Christian and *Niamtre* Khasi boys. Basu and Sun (2004a) studied physical growth and development among the Khasi adolescent children of Meghalaya. They (Basu and Sun, 2004b) also studied the Khasi adolescent girls for the circumference of the upper arm and triceps skinfolds thickness. The Khasi girls stand far apart from the International Reference (NCHS) is what the study reveals. Studies based on the NFHS-II shows that Meghalaya had the highest percentage of stunted (44.60%) and children with wasting

(14.00%) in comparison with children of other Northeastern states (Rao et al., 2004). Height and sitting height of Khasi girls aged 3 to 18 years were also measured (Khongsdier and Mukherjee, 2005) to test the effect of heterosis. They recorded hybrid girls (Khasi Muslim) were larger than the non-hybrid Khasi girls across age. Wariri (2009) studied the growth and nutritional status of Khasi children of Meghalaya. Physical growth and nutritional status of the Pnar adolescents of Jaintia hills of Meghalaya is studied by Kamar (2011).

In Mizoram, Urade and Mukherjee (2002) studied growth and development of the Lusai children of Aizawl district. Varte (2003) also studied the health and nutritional status of the Hmars of Aizawl district. Khongsdier et al. (2005) studied the Hmar of Mizoram and Khasi of Meghalaya to determine the sex difference in Chronic Energy Deficiency among adolescents in the context of patrilineal and matrilineal societies in Northeast India. The sex differences in the prevalence of CED were 5.00 percent, 10.00 percent, and 6.00 percent in the Khynriam Khasis, War Khasis, and Hmars respectively.

Jamir (2001) studied the growth and nutritional status of the Ao Nagas of Nagaland. Later, Tsopoe (2003) studied the nutritional status and physical growth of Lotha Naga children of Wokha district of Nagaland. They are relatively taller and heavier than the Khasi children of Meghalaya and Meitei children of Manipur. The nutritional status of the children is moderate in comparison with other population groups of Northeast India. Longkumer (2013) studied nutritional status and somatotypes of Ao children of Nagaland.

It is evident from the above overview that the literature on growth and nutritional studies on Northeast Indian population are still inadequate in number in relation to the area with a large and varied nature of ethnic inhabitants. Auxological studies in Northeast India are relative of recent origin as it started in the sixties of the nineteenth century and are mainly confined to the state of Assam and Meghalaya. A few numbers of fragmentary reports on growth and nutritional status of the people of the region have appeared in the last five-six decades, however, almost all the studies were based on cross-sectional data and the majority of them are concerned with physiological changes and to assess existing variation between populations. There is a dearth of information on the relationship between body composition/dimensions and morbidity patterns. Very limited studies were done on the role of socio-economic factors influencing the growth form. Rao et al., (2004) based on NFHS-II data reported that nutritional status of female children of Northeast India has a nutritional edge over male children. In the Northeast region of India, Assam and Meghalaya are the two states which record the highest prevalence of undernutrition (NFHS-II, 2000; NFHS-III, 2007). There are still many areas that urgently need to carry intensive research relative to health and nutritional status at the population level based on extensive fieldwork and observation in this part of the country, especially with regard to the emerging trend of the double burden of under and over nutrition.

In the present compilation, specialists in human growth and nutritional studies present a total of thirteen well-postulated research articles to acquaint the readers a number of extensive studies on growth and nutritional issues of several ethnic groups of Northeast India. The compilation features with an abundance of materials and wide-ranging analysis which is extremely thought-provoking, useful and will be accepted as a valuable collection to enhance the understanding and serve the interest of biological anthropologists, nutritionists, pediatricians, besides other social scientists, policy makers and anyone with a concern about the population groups of Northeast India.

Reference

Ashizawa, K, and Kawabata, M 1984, 'A cross-sectional anthropometric survey among the Meitei girls of Thoubal District, Manipur'. *Annals of Human Biology*, Vol. 17, pp. 437-443.

Basu, S, and Sun, D 2004a, *Adolescence Growth Pattern among the Khasi of Meghalaya*, Unpublished Report Submitted to Anthropological Survey of India, Shillong.

Basu, S, and Sun, D 2004b, *A study on muscle and fat status among the Khasi tribal adolescent girls of Meghalaya, India*. Presented in the Congress of International Union of Anthropological and Ethnological Sciences held in Kolkata from 12th to 15th December 2004.

Begum, G 1995, *Age Changes in Some Somatometric Characters with Reference to Nutritional Status of Assamese Muslims of Kamrup District, Assam*. Ph.D. Thesis, Gauhati University.

Begum, G 1998. 'Nutrition and economic status in relation to change in adult growth among the Assamese Muslim of Kamrup District. Assam'. *Journal of Human Ecology*, Vol. 9, pp. 521-523.

Brown, KH 1984, 'Measurement of dietary intake'. *Population Development Review. (Supplementary),* Vol. 10, pp. 69-91.

Begum, G and Choudhury, B 1997, 'Socio-economic interactions with growth and nutritional status among the Assamese Muslims of Kamrup district, Assam'. *Indian Journal of Physical Anthropology and Human Genetics,* Vol. 20, pp. 119-146.

Begum, G and Choudhury, B 1999, 'Age changes in some somatometric characters of the Assamese Muslims of Kamrup district, Assam'. *Annals of Human Biology*, Vol. 26, pp. 203-217.

Chelleng, PK and Mahanta, J 1998, 'Growth and development of tea garden children of Assam (1-12 years)'. *Journal of Human Ecology,* Vol. 9, pp. 399-401.

Choudhury, B 1979, *A Cross-Sectional Study of Growth of Some Somatometric Characters of Rabha Boys of Kamrup District, Assam.* Ph.D. Thesis, Gauhati University.

Choudhury, B and Gogoi, M 1995, 'Pre-adolescent growth of neo Assamese Muslim children of Goraimari Block, Kamrup district'. *Bulletin of the Department of Anthropology, Gauhati University,* Vol. 9, pp. 87 - 112.

Choudhury, B and Das, PB 1985, 'A mixed longitudinal study on growth in height and weight of Assamese children from 1-7 years'. In B. M. Das (Ed): *Children of North East India.* pp. 37- 42, Anthropological Survey of India, Calcutta.

Choudhury, B and Bhuyan, SC 1994, 'Growth and nutritional status of Kalita children of Kamrup District, Assam'. *Bulletin of the Department of Anthropology, Gauhati University*, Vol. 9, pp. 122-134.

Choudhury, B, Begum, G and Barua, T 1992, 'Growth pattern of school children of Gauhati city'. *Bulletin of the Department of Anthropology, Gauhati University*, Vol. 6, pp. 74 - 86.

Das, AK 1982, *Growth and Body Build of the Assamese and Rajasthani Boys Aged 6 to 17 Years: A Comparative Study*. Ph.D. Thesis. Dibrugarh University, Dibrugarh.

Das, PB 1966, *A Study on Growth of Somatometric Characteristics of a Selected Caste of Assam*. Ph. D. Thesis, Gauhati University.

Das, BM and Das, PB 1969-71, 'A study on some aspects of the growth of the Assamese boys'. *Journal of Gauhati University (Science)*, Vol. 20-22, pp. 51-65.

Das, PB 1971, *Growth studies of Kalita Children*. Field Research Project, Miami, USA.

Das, PB 1973a, 'A cross-sectional study of growth in height and weight of an endogamous caste of Assam'. *Bulletin of the Department of Anthropology Gauhati University*, Vol. 2, pp. 24-31.

Das, PB 1973b, 'Height, weight and chest circumference of Vaishya children'. *Journal of Assam Science Society*, Vol. 16, pp. 34-42.

Das, PB 1974, 'A comparative study of growth in respect of height, weight and chest circumference of rural Assamese boys'. *Bulletin of the Department of Anthropology, Dibrugarh University*. Vol. 3, pp. 43-55.

Das, PB 1977, 'A longitudinal study of growth in height, weight and sitting height of five siblings for a period of ten years'. *Abstract of the Proceedings of 64th Annual Session of Indian Science Congress Association*.

Das, PB 1978, 'A longitudinal study on the growth of head and face dimensions of four sisters for twelve years'. *Proceedings of the Xth International Congress of Anthropological and Ethnological Sciences,* New Delhi.

Das, B 2001, *Growth and Development of the Kaibarta Children of Assam: Bio-cultural Perspective*. Unpublished Report Submitted to Anthropological Survey of India, Shillong.

Das, H 1988, *Age Changes in Some Somatometric Characters of the Hajongs of Meghalaya.* Ph.D. Thesis, Gauhati University.

Das, B, and Boruah, G 2002, *Growth and Development of the Ahom Children of Assam: Bio-cultural Perspective.* Unpublished Report Submitted to Anthropological Survey of India, Shillong.

Das, PB and Choudhury, B 1982, 'A mixed longitudinal studies on growth in height and weight of Assamese children from 1 to 7 years'. *Bulletin of the Anthropological Survey of India, Vol.* 31, no. (3-4), pp. 37-42.

Das, PB and Choudhury, B 1992, 'A mixed longitudinal study on growth in height and weight of Assamese children from 8-16 years'. *Bulletin of the Department of Anthropology, Gauhati University,* Vol. 6, pp. 61-69.

Das, PB and Hazarika, J 1975, 'Height and weight of post-graduate students of Gauhati University'. *Journal of Assam Science Society,* Vol. 18. no.1. pp. 11-17.

Deori, J 2016, *A cross-sectional study on physical growth among the Deoris of Assam and Arunachal Pradesh*. Ph.D. Thesis, Gauhati University.

Devi, S 1986, 'A comparative study of growth with regard to height and weight between the Meitei and Muslim girls of Manipur'. *Bulletin of the Department of Anthropology, Dibrugarh University,* Vol. 15, pp. 140-150.

Duarah, DK 1985, *A Study on the Physical Anthropology of the Monpas of Arunachal Pradesh with Special Reference to the Growth of Monpa Boys.* Ph.D. Thesis, Gauhati University.

Duarah, DK and Das, PB 1978, 'A cross-sectional study of growth in four somatometric characters of the Khampti boys of Arunachal Pradesh'. *Resarun*, Vol. 10, pp. 15 - 24.

Dutta, MN 1997, *A Comparative Study of Growth Pattern in the Early Childhood among the Children of Two Ethnic Groups of Dibrugarh, Assam.* Ph.D. Thesis, Dibrugarh University.

Dutta, D 2012, *A Study on Growth and Nutritional Status among the Ahoms of Upper Assam.* Ph.D. Thesis, Dibrugarh University.

Gaur, R. and Singh, NY 1995, 'Pattern of growth among rural school going children of Manipur'. *Man in India*, Vol. 75, no.3, pp. 269-281.

Haokip, JV 2015, *Growth, Body Dimensions, and Nutritional Status among the Thadou Kukis of Chandel district, Manipur.* Ph.D. Thesis, North Eastern Hill University.

Hazarika, ANZA 1978, A *Study of Growth of Head, Face, Trunk, and Limbs of the Ahom Boys of Assam.* Ph.D. Thesis, Dibrugarh University.

Hussain, A 2003, *Somatotypes, Physical Growth and Nutritional Status of the Bengali Muslim Boys of Barak Valley, Assam: A Bio-anthropological Study.* Ph.D. Thesis, Gauhati University.

International Institute for Population Sciences 2000, National Family Health Survey (NFHS-2), 1998-99, India. Mumbai: International Institute for Population Sciences; p. 438.

International Institute for Population Sciences 2007, National Family Health Survey (NFHS-3), 2005-06, India. V. II. Mumbai: International Institute for Population Sciences; p. 168.

Jamir, TA 2001, *Nutritional status and Physical growth of Ao children of Mokokchung District, Nagaland.* Ph.D. Thesis, North Eastern Hill University.

Jibonkumar, H and Singh, TS 2006, 'Nutritional status of the Sekmai children, Manipur'. *International Journal of Anthropology, Vol.* 21, pp. 183-191.

Kamar, P 2011, *Physical Growth and Nutritional Status among the Pnar adolescents of Jaintia Hills, Meghalaya.* Ph.D. Thesis, North Eastern Hill University.

Khongsdier, R 1995, *A Study on Nutrition and Health Status in Relation to Some Bio Social Factors among the War Khasi of Meghalaya.* Ph.D. Thesis, North Eastern Hill University.

Khongsdier, R 1996, 'Assessment of growth and nutritional status: An anthropological perspective'. *Acta Medica Auxologica*, Vol. 28, pp. 147-153.

Khongsdier, R 2001, 'Body Mass Index of adult males in 12 populations of Northeast India'. *Annals of Human Biology,* Vol. 28, no. 4, pp. 374-383.

Khongsdier, R 2002, 'Body Mass Index, and morbidity in adult males of the War Khasi in Northeast India'. *European Journal of Clinical Nutrition,* Vol. 56, pp. 484-489.

Khongsdier, R and Mukherjee, N 2003, 'Growth and nutritional status of Khasi boys in North East India relating to exogamous marriages and socio-economic classes'. *American Journal of Physical Anthropology*, Vol. 122, pp. 162-170.

Khongsdier, R, Varte, R and Mukherjee, N 2005, 'Excess male chronic energy deficiency among adolescents: A cross-sectional study in the context of patrilineal and matrilineal societies in North East India', *European Journal of Clinical Nutrition,* Vol. 59, pp. 1007-1014.

Kotal, M, and Boruah, P 2004, *Physical Growth of Adolescents among the Sonowal Kacharis of Dibrugarh District, Assam.* Unpublished Field Research Project Submitted to Anthropological Survey of India, Shillong.

Kulkarni, S and Sarma, ER 2001, 'Nutrition among Karbi tribe of Assam: An anthropometric assessment'. *Journal of Indian Anthropological Society*, Vol. 36, pp. 223-229.

Longkumer, T 2013, *A Study on Somatotype and Nutritional Status in relation to Physical activities among the Ao Children of Mokokchung District, Nagaland.* Ph.D. thesis, North Eastern Hill University.

Luwang, NC 1981, 'A comparative study of weight pattern in children under five years of age, hill and valley areas of Manipur'. *Indian Pediatrics*, Vol. 18, pp. 283-286.

Mantaw, NH and Priyadarshini, P 2000, 'Nutritional status of children in Arunachal Pradesh: An overview'. *Resarun*, Vol. 26, pp. 109-118.

Medhi, D 2013, *Growth and Nutritional status among the Muslims of Barpeta District Assam.* Ph. D. Thesis, Gauhati University.

Medhi, GK, Barua, A and Mahanta, J 2006, 'Growth and nutritional status of school-age children (6-14 years) of tea garden workers of Assam', *Journal of Human Ecology*, Vol. 19, no. 2, pp. 83-85.

Mukherjee, N 2002, *Studies on Demography and Growth Pattern among the Khasi Children of Shillong, Meghalaya.* Ph.D. Thesis, North Eastern Hill University.

Pathak, RK 1995, 'Stature, weight, and menarche in adolescent Khasi girls of Shillong', *Journal of Anthropological Society of Manipur*, Vol. 2, pp. 108-113.

Rao, GR, Ladusingh, L and Pritamjit, R 2004, 'Nutritional Status of Children in North East India'. *Asia-Pacific Population Journal*, Vol. 19, pp. 39-56.

Sahani, R, and Das, B 2003, *Growth of Adolescents of Ahom Children of Sibsagar District, Assam*. Unpublished Field Research Project. Submitted to Anthropological Survey of India, Shillong.

Sikdar, M 2010, *Bio Social Determinants of Growth and Nutritional Status of the Mishings of Upper Assam*. Ph. D. Thesis, Dibrugarh University.

Sikdar, M and Ahmed Das F 2006, 'Growth progression with ages in physical and physiological variables among the Bengali population of Dibrugarh, Assam'. *Bulletin of Department of Anthropology Dibrugarh University*, vol 34, pp. 45-49.

Singh, LR 1986, *Physical Variation between Two Sections of the Kabuis of Manipur and Their Ethnic Position.* Ph.D. Thesis, Gauhati University.

Singh, J 2008, *A Study on Growth and Nutritional Status of the Sonowal Kachari of Dibrugarh District, Assam*. Ph.D. Thesis, Dibrugarh University, Dibrugarh.

Singh, LD and Singh, TS 2000, 'Physical growth among the affluent and non-affluent Meitei boys of Manipur'. *Man in India,* Vol. 80, no.3- 4, pp. 295-307.

Singh, NY and Konsam, S 2001, 'A study on physical growth and nutritional status among school children of Thanga Island, Manipur'. In Konsam S. Singh (Ed): *Glimpses on the Culture and Biology of the People of North East India.*, pp 148-174. Har Anand Publications, New Delhi.

Singh, L, Shyamacharan, T and Singh Rajendra, L 2001, 'Skinfold thickness of the Meitei boys of Manipur'. *Journal of Human Ecology*, Vol. 12, no. 2, pp. 115-119.

Talwar, I and Singh, AB 1995, 'Growth pattern of adolescent Meitei boys and girls of Manipur', *Man in India,* Vol. 75, no. 3, pp. 231-240.

Tsopoe, T 2003, *Nutritional Status and Physical Growth of Lotha Children of Wokha District, Nagaland.* Ph.D. thesis, North Eastern Hill University.

Urade, BP, and Mukherjee, SK 2002, *Growth and Development of the Lushai Children of Mizoram: Bio-cultural Perspective*. Unpublished Project Report Submitted to Anthropological Survey of India, Kolkata.

Varte, LP 2003, *Health Status of the Hmars of Aizawl District, Mizoram*. Unpublished Report Submitted to Anthropological Survey of India, North Eastern Regional Centre, Shillong.

Wariri, BC 2009, *Growth and nutritional status of the Khasi children in West Khasi Hills District of Meghalaya*. Ph.D. thesis, North Eastern Hill University.

Acknowledgement

Our sincere gratitude goes to all the contributors and those who suggested and encouraged for this project. We would like to acknowledge for their academic endeavor as well as moral support towards this project.

Our cordial thanks go to the dedicated and hardworking staff of Gyan Publishing House, New Delhi, we have dealt with personally and the others behind the scenes for providing a smooth transition to reach our ultimate goal. We would also like especially to thank Mr. Manimugdha Bhuyan of the Department of Anthropology, Dibrugarh University for arranging photographs for the jacket.

Last but most important; our special thanks to our parents and all the family members for their encouragement and support throughout the endeavor which make our effort a successful one.

Dibrugarh
19 February 2019

Sarthak Sengupta
Dali Dutta

Acknowledgement

List of Contributors

A.F. Gulenur Islam Barbhuya, Department of Anthropology, M.H.C.M. Sc. College, Hailakandi, Assam.

Gulrukh Begum, Department of Anthropology, Gauhati University, Bordoloi Nagar, Guwahati, Assam, 781014.

Prema Bharati, Biological Anthropology Unit, Indian Statistical Institute, 203 Barrackpore Trunk Road, Kolkata, 700108.

Susmita Bharati, Biological Anthropology Unit, Indian Statistical Institute, 203 Barrackpore Trunk Road, Kolkata, 700108.

Juri Borah, Department of Anthropology, Dibrugarh University, Dibrugarh, Assam, 786004.

Kaushik Bose, Department of Anthropology, Vidyasagar University Midnapore-721 102, West Bengal, India.

Kenneth Darnei, Department of Anthropology, North Eastern Hill University, Shillong, Meghalaya, 793022.

Rekha Das, Department of Anthropology, Gauhati University, Bordoloi Nagar, Guwahati, Assam, 781014.

R. K. Neeta Devi, Department of Anthropology, Umshyrpi, College Laban, Shillong.

Johnsly Wells Dkhar, Department of Anthropology, North Eastern Hill University, Shillong, Meghalaya, 793022.

Dali Dutta, Department of Anthropology, Dibrugarh University, Dibrugarh, Assam, 786004.

S. Khamu, Department of Anthropology, Northeastern Hill University, Shillong, Meghalaya, 793022.

B. T. Langstieh, Department of Anthropology, North Eastern Hill University, Shillong, Meghalaya, 793022.

Temsumongla Longkumer, 338 Parkview dr, Phoenixville, PA 19460, USA.

Temsutola Maken, 2/4 Block 2 New Minto Road Hostel, Minto Road. New Delhi, 110002.

Nitish Mondal, Department of Anthropology, Assam University (Diphu Campus), Diphu, Karbi Anglong-782462, Assam, India.

Manoranjan Pal, Economic Research Unit, Indian Statistical Institute, 203 Barrackpore Trunk Road, Kolkata, 700108.

Flora Rabha, Department of Anthropology, Gauhati University, Bordoloi Nagar, Guwahati, Assam, 781014.

Harbamon Rongpi, Department of Anthropology, Assam University (Diphu Campus), Diphu, Karbi Anglong-782462, Assam, India.

Sarthak Sengupta, Department of Anthropology, Dibrugarh University, Dibrugarh, Assam, 786004.

Maibam Samson Singh, Department of Anthropology, Sikkim University, Gangtok, Sikkim.

L. Robert Varte, Sc 'D', Defense Institute of Physiology and Allied Sciences, Defense Research and Development Organization, Lucknow Road, Timarpur, Delhi, 110054.

List of Tables

List of Figures

Chapter - 1

Overweight and Obesity among the Adult Population in Northeast India

— Premananda Bharati, Manoranjan Pal and, Susmita Bharati

Introduction

In the last fifteen to twenty years, an increase in obesity is very much alarming. It is estimated worldwide that among individuals aged 15 years and above, about 2.3 billion people will be overweight and over 700 million will be obese in 2015 (WHO, 2013). According to WHO (2013) that overweight and obesity are the fifth leading risk of deaths every year and also cause the different burden of diseases like diabetes (44.0%), ischemic heart disease (23.0%) and certain cancer burdens (7.0% to 41.0%). Obesity is not a single disorder and it is dependent on multi-factorial disorder including socio-cultural and environmental factors such as nutrition, stress exercise etc. On the other hand, it is anticipated that inheritance is also responsible for obesity though the genetic mechanism is not known. The existence of the genetic effect is expressed by environment and behavioral factors (Pi-Sunyer, 1994).

Today overweight or obesity is not only restricted in developed countries but it is also the problem of developing countries. The developing countries are also facing the problems of dual burden of malnutrition i.e. facing at a time problem of under as well as overnutrition. Many scholars explained this as due to 'nutritional transition' in developing countries.

More than 30 million obese people are residing in India and the number is increasing alarmingly (NFHS-2; NFHS-3). The problem is more acute among women than men and mainly in urban India where 23 percent of women are either overweight or obese and among men, it is 20 percent (NFHS-3). Obesity is causing due to rapid and unplanned urbanization, intra-urban socio-economic disparity and health inequality in an urban area.

It is also to be noted that like any other developing countries, in India, obesity is directly related to higher socio-economic status. Thus, it is more seen in urban areas.

Here our focus of the study is the condition of overnutrition among the adults in Northeast India. Northeast India refers to the easternmost region of India, consisting of eight different states. It is accounted for 7-9 percent of the total land space of the country. It is characterized by low per capita income, low capital formation, inadequate infrastructural facilities, geographic isolation, and communication bottleneck. Here the tribal people are more in numbers. It is reported that the status of women is comparatively better in tribal society (Handsdak, 2012; Roy Burman, 2012). The scheduled tribes are 94.75 percent in Mizoram, 87.70 percent in Nagaland, 85.53 percent in Meghalaya, 63.66 percent in Arunachal Pradesh, 34.41 percent in Manipur, 30.95 percent in Tripura, and 12.82 percent in Assam (Zehol, 2010). Except for Assam and Tripura, the rest six states are predominated by tribal population. In such a background of the Northeastern region, the objective of the present study is (i) to see the overweight or obesity among the adult population including both male and female of Northeast India, and (ii) to see the relationship between obesity with different socio-economic variables.

Material and Methods

In the present study, we have used the National Family Health Survey (NFHS –III) data conducted by the International Institute for Population Sciences (IIPS), Mumbai, in 2005-06. They collected data on reproductive-aged women of 15-49 years and 15-54 years men from 29 states in India and the population frame was based on the 2001 census. The data are cross-sectional and unit-level data. As our focused area of study is on adult women and men in Northeast India, we have considered the data of eight states of Northeast India. However, we have considered women of age-group 20-49 years and men in the age group 20-54 years. The sample size consists of 17,004 for women and 10,030 for men. Body Mass Index (BMI) is considered as a basis for measuring obesity. It is classified into the following categories (WHO, 2003) as follows:

Categories	Ranges of BMI (kg/m^2)
Chronic Energy Deficiency (CED)	<18.5
Normal	18.5 – 24.9
Overweight	25.0 – 29.9
Obese	30.0 only.

The background explanatory variables are (i) age-groups (ii) type of place (iii) education (iv) occupation (v) diabetes (vi) Asthma (vii) Thyroid and (viii) wealth index of the family.

To determine the differences in the prevalence of obesity between male and female for each state, the percentage of overweight or obese men was subtracted from that of women. Thus, a positive sign means the percentage is more among women and a negative sign means just the opposite trend. Regression analysis was performed to establish the relationship between a dependent (overweight or obese) and independent (socio-economic) variables. The risk of obesity was regressed on various independent variables using categorical logistic regression analysis. Dependent variables are taken as binary. Overweight or obesity was considered as '1' and others are coded as '0'. An estimated Odd ratio of '1' indicates that the nature of the dependent variable is not different from that of the reference category. If the estimated Odd ratio is >1, the probability of becoming affected is more in this category compared to the reference category and if it is <1, then it is just opposite to that of '>1' case.

Results

Table 1.1 shows the percentage distribution of overweight or obesity among adult female and male in Northeast India as a whole and in different states in Northeast India. Percentage of overweight or obesity among female and male are 12.5 and 10.1 respectively. However, the percentages vary over different states. The highest percentage of obesity among women and men are found in Sikkim (19.6%; 15.6%) respectively and the lowest are

Table 1.1: Percentage distribution of overweight and obesity among adult men and women of Northeast India

Zones and states of India	*Women*				*Men*				*Differences in obesity between men and women*
	N	*Overweight*	*Obese*	*Overweight and obese*	*N*	*Overweight*	*Obese*	*Overweight and obese*	
Northeast India	17004	10.7	1.8	12.5	10030	9.3	0.8	10.1	2.4
Arunachal Pradesh	1236	9.6	1.4	11.0	55	9.5	0.0	9.5	1.5
Assam	2995	9.7	1.5	11.2	1019	6.5	0.5	7.0	4.2
Manipur	3624	14.2	2.4	16.6	3108	10.6	1.2	11.8	4.8
Meghalaya	1524	6.4	1.0	7.4	481	7.5	0.2	7.7	-0.3
Mizoram	1477	10.1	1.5	11.6	546	12.8	0.5	13.4	-1.8
Nagaland	3043	9.2	1.2	10.4	3082	8.0	0.8	8.9	1.5
Sikkim	1643	15.6	4.0	19.6	653	14.1	1.5	15.6	4.0
Tripura	1462	8.1	1.0	9.1	586	6.5	0.0	6.5	2.6

found in Meghalaya for women (7.4 %) and Tripura for men (6.5 %). The magnitude of obesity is more among women than men. This trend is the same in most of the states in Northeast except in Meghalaya and Mizoram.

Table 1.2 presents the relationship between overweight or obesity with different socio-economic variables. It is seen that obesity is directly related to age, education, white coloured jobs and wealth index of the family and these results are statistically significant at 1 percent level of significance. It is also seen that obesity is directly related to urban areas and diabetes. And these results are also statistically significant at 1 percent level of significance.

Table 1.2 Socio-demographic conditions in different states of Northeast India

Socio-economic Variables	*Arunachal Pradesh*		*Assam*		*Manipur*		*Meghalaya*		*Mizoram*		*Nagaland*		*Sikkim*		*Tripura*	
	F	*M*	*F*	*M*	*F*	*M*	*F*	*M*	*F*	*M*	*F*	*M*	*F*	*M*	*F*	*M*
Age groups																
20-29	46.9	40.4	45.0	36.5	42.9	38.5	48.2	42.8	43.1	41.6	48.1	40.4	46.4	45.6	40.2	35.3
30-39	35.9	30.8	35.4	32.1	35.1	30.7	31.2	30.4	34.7	29.7	33.3	30.6	53.7	27.6	34.0	30.0
40 or more	17.2	28.8	19.6	31.4	22.1	30.8	20.6	26.8	22.3	28.8	18.5	29.0	19.9	26.8	25.8	34.6
Place of residence																
Rural	69.1	70.1	68.0	66.9	54.9	56.1	57.0	58.2	45.8	47.8	50.6	47.4	60.7	59.7	74.4	76.1
Urban	30.9	29.9	32.0	33.1	45.1	43.9	43.0	41.8	54.2	52.2	49.4	52.6	39.3	40.3	25.6	23.9
Education																
Illiterate and primary	63.4	45.6	47.7	36.6	34.3	16.1	45.8	40.5	28.4	22.2	37.4	30.2	46.1	36.9	50.1	43.0
Secondary	30.0	43.0	43.7	50.1	47.4	58.3	41.7	46.2	63.2	68.3	52.4	53.6	44.2	49.0	43.7	47.1
Higher	6.6	11.4	8.6	13.2	18.3	9.5	12.5	13.3	8.5	9.5	10.3	16.2	9.7	14.1	6.2	9.9
Occupation																
Not working	42.3	39.9	4.3	19.0	23.4	32.9	16.7	10.4	23.4	33.8	23.8	25.7	10.3	28.2	12.2	26.0
Professionals	37.1	7.5	69.6	7.9	32.5	10.3	55.1	11.1	50.3	6.2	53.7	12.6	64.6	9.0	65.8	5.0
Service/sales	5.5	19.7	14.8	34.7	23.1	24.3	8.5	37.6	2.8	30.5	2.5	20.5	3.3	24.2	8.5	35.0
Agriculture	9.6	18.8	5.5	28.9	13.8	20.7	13.5	33.4	16.6	20.4	10.8	28.0	13.9	26.6	6.8	20.1
Domestic/manual	5.6	14.1	5.8	9.5	7.2	11.8	6.2	7.5	6.9	9.2	9.2	13.2	7.9	11.9	6.7	13.9
Having Diabetes																
No	99.7	97.7	99.5	98.5	98.8	98.8	98.6	98.5	98.7	97.1	99.2	97.6	97.6	98.0	97.9	94.4
Yes	0.3	2.3	0.5	1.5	1.2	1.2	1.4	1.5	1.3	2.9	0.8	2.4	2.4	2.0	2.1	5.6
Having Asthma																
No	98.0	97.7	98.5	98.5	98.5	98.8	98.0	98.5	96.1	97.1	98.7	97.6	95.0	98.0	93.6	94.4
Yes	2.0	2.3	1.5	1.5	1.5	1.2	2.0	1.5	3.9	2.9	1.3	2.4	5.0	2.0	6.4	5.6
Having Thyroid problems																
No	97.7	99.6	99.3	99.0	97.2	98.4	98.8	99.6	97.9	99.5	99.5	99.2	98.7	99.4	97.8	99.5
Yes	2.3	0.4	0.7	1.0	2.8	1.6	1.2	0.4	2.1	0.5	0.5	0.8	1.3	0.6	2.2	0.5
Wealth Index																
Poorest and poorer	40.0	42.3	39.9	39.3	14.3	14.5	25.1	25.6	7.7	7.5	20.2	21.0	7.2	7.5	29.3	26.3
Middle	21.4	20.7	22.1	22.4	29.9	31.6	22.2	21.8	16.9	15.4	24.1	23.9	16.5	19.3	42.1	46.8
Richer and Richest	38.7	36.9	38.0	38.4	55.8	53.9	52.6	52.6	75.5	77.1	55.8	55.8	76.3	73.2	28.6	27.0

Table 1.3: Incidence of obesity among adult men and women of Northeast India in relation to different socio-economic variables

Independent variables	*Women*				*Men*			
	N	*Others*	*Overweight and obese*	*Chi-square*	*N*	*Others*	*Overweight and obese*	*Chi-square*
Age-groups (in years)								
20-29	7667	93.4	6.6	483.739	3976	94.8	5.2	183.448
30-39	5826	84.2	15.8	df = 2	3059	88.0	12.0	df = 2
40 or more	3511	80.0	20.0	Sig. = 0.000	2995	85.4	14.6	Sig. = 0.000
Place of residence								
Rural	10050	91.7	8.3	392.420	5653	93.5	6.5	182.055
Urban	6954	81.4	18.6	df = 1 Sig. = 0.000	4377	85.3	14.7	df = 1 Sig. = 0.000
Education								
Illiterate and primary	7197	91.2	8.8	183.46	2864	94.7	5.3	265.056
Secondary	7924	85.6	14.4	df = 2	5392	90.6	9.4	df = 2
Higher	1882	81.2	18.8	Sig. = 0.000	1769	80.0	20.0	Sig. = 0.000
Occupation				248.080				
Not working	8945	87.1	12.9	df = 4	1005	94.3	5.7	
Professional	1207	79.2	20.8	Sig. =	1193	77.8	22.2	393.044
Service/ sales	1887	81.8	18.2	0.000	2463	85.1	14.9	df = 4
Agriculture related	3163	93.5	6.5		2791	95.7	4.3	Sig. =
Skilled/unskilled/manual	1771	90.3	9.7		2553	92.2	7.8	0.000
				\				
Having Diabetes				66.685				40.173
No	16821	87.7	12.3	df = 1	9863	90.1	9.9	df = 1
Yes	182	67.6	32.4	Sig. = 0.000	165	75.2	24.8	Sig. = 0.000
Having Asthma								0.002
No	16576	87.6	12.4	5.935	9819	89.9	10.1	df = 1
Yes	428	83.6	16.4	df = 1 Sig. = 0.009	210	90.0	10.0	Sig. = 0.988
Having Thyroid problem				5.027				0.090
No	16730	87.6	12.4	df =12	9929	89.9	10.1	df = 1
Yes	271	83.0	17.0	Sig. = 0.016	100	89.0	11.0	Sig. = 0.738
Wealth index								
Poorest and poorer	3884	96.6	3.4	750.115	2100	97.6	2.4	403.640
Middle	4215	92.9	7.1	df = 2	2651	95.1	4.9	df = 2
Richer and richest	8925	81.0	19.0	Sig. = 0.000	5279	84.2	15.8	Sig. = 0.000

Table 1.3 describes the socio-demographic conditions in different states of Northeast India. It is seen that in all the states of Northeast India, the average range of rural people are 45 to 75 percent, though the highest percentage of rural people is found in Tripura and the low percentages are

found in Manipur, Mizoram, and Meghalaya. The highest percentage of illiterate males and female are found in Arunachal Pradesh. The highest 'higher educated women' are found in Manipur and the highest 'higher educated men' are found in Mizoram. Regarding economic conditions, the highest 'poorer or poorest' economic conditions are found in Arunachal Pradesh, followed by Assam then Tripura and relatively high 'richer or richest' wealth index are found in Sikkim, Mizoram, Manipur, and Meghalaya.

Table 1.4 shows the result of categorical logistic regression of different socio-economic variables with obesity of women and men in Northeast India. The results show that obesity is increasing among both females and males with the increment of age, education upliftment, and maximum obesity are noticed high among professional jobholders followed by service holders, and better economic condition families. And these results are statistically significant at 1.0 percent level of significance. To find the relationship between obesity with some non-communicable diseases, diabetes has a significant relationship with obesity.

Table 1.4: Categorical Logistic regressions of obesity among adult men and women of Northeast India in relation to different socio-economic variables

Independent variables	***Women***	***Men***
	Odd ratios	***Odd ratios***
Age-groups		
20-30®	1.00	1.00
31-40	2.809**	2.236**
41 or more	3.822**	2.846**
Place of residence		
Rural®	1.00	1.00
Urban	1.444**	1.377*
Education		
Illiterate and primary®	1.00	1.00
Secondary	1.135*	1.144
Higher	1.118	1.791**
Occupation		
Agriculture related®	1.00	1.00
Not working	1.348**	0.896
Skilled/unskilled/manual	1.200	1.393**
Service/sales	1.474**	1.667**
Professional	1.616**	2.043**

(Contd...)

Having Diabetes		
No®	1.00	1.00
Yes	2.084**	1.746**
Having Asthma		
No®	1.00	1.00
Yes	1.204	1.095
Having Thyroid problem		
No®	1.00	1.00
Yes	1.045	0.785
Wealth index		
Poorest and poorer ®	1.00	1.00
Middle	1.981**	1.818**
Richer and richest	4.732**	4.507**

® = Reference category

Discussion

The study is an attempt to look at the occurrence of overweight or obesity among adult Indians in Northeast India. It tries to answer the question of the prevalence of obesity in different states of Northeast India. It is seen from our study that the highest obesity is found in Sikkim among both women and men. The reason is that in Sikkim, in general, the percentage of poor category people is very less (among female only 7.2 % and among male, it is 7.5%). Among females, the lowest occurrences of obesity are found in 7.4 percent in Meghalaya while among male, it is found in Tripura as 6.5 percent. It is also seen that there lie gender differences in overweight or obesity among adult Indian people of Northeast India because 12.5 percent of adult women and 10.1 percent of the adult male are overweight or obese. So occurrences of overweight or obesity are more among women than men. The magnitude of the difference is 2.4. This trend is similar in six states of Northeast India except in Meghalaya, and Mizoram. In Meghalaya, among women, the percentage of poorer women is nearly 25.1 percent and the lowest percentage of obesity is found also in Meghalaya. So, this finding proves that in developing countries, the economy is directly related to obesity. In Tripura, the percentage of poor males are 26.3 percent and obesity are the lowest while in Assam and Arunachal Pradesh, 40.0 percent men belong to poorer classes and percentage of obesity is not more than that of Tripura. So, it is highly correlated with the economic condition of the family and also proves that in developing countries, the economy is directly related to obesity (Sobol and Stunkard, 1989). To find the correlation with other factors, like education, the highest percentage of illiterate persons is found in Arunachal Pradesh and a relatively high percentage of higher educated females and males are found in Manipur and Nagaland respectively. It is also noticeable that out of eight states in Northeast India, in six states (barring Meghalaya and Mizoram), females are more obese than males. This observation supports the view that women had a consistently lower level of recommended activity than men (Baldridge, 2002). It is also known that the causes of overweight or obesity among women have been found due to endocrine hormonal factor but more probably it may be due to socio-cultural factors also. In Meghalaya and Mizoram, relatively higher percentages of professionals and service holders are found. Regarding education, the maximum percentage of secondary and higher educated persons are also found in these two states. These results support the view of Ingle and Nath (2008), who opines that middle and higher income groups are prone to develop obesity due to a sedentary lifestyle and less physical activity due to white coloured job status.

Therefore, from the above discussion it can be concluded that besides biological factors, obesity is associated with some socio-economic factors and urbanization. As development is synonymous with urbanization, therefore, appropriate precautionary measures must be taken to prevent overweight. And if this trend continues in the near future overweight will be one of the serious public health problems among adults.

References

Baldridge, D 2002, *Behavioral risk factor surveillance system document*, CDC.

Handsdak, EV 2012, The status of tribal women. Fr. Agapit Turkey (ed.) *Reporting to India's social challenges/- promoting Tribal Rights and culture* (online) Available: httpHYPERLINK "http:///":// www. Holycross justice .org/pdf/ Asia/ Indias social challenges/ promoting Tribal Rights/The status of tribal women. pdf.

Ingle, GK, and Nath, A 2008,' Geriatric health in India: concern and solutions', *Indian Journal of Community Medicine,* Vol. 33, pp. 214-218.

International Institute for Population Sciences 2000, National Family Health Survey (NFHS-2), 1998-99, India. Mumbai: International Institute for Population Sciences; p. 438.

International Institute for Population Sciences 2005-06, National Family Health Survey (NFHS-3), India. V. I. Mumbai: International Institute for Population Sciences; 2007. p. 540.

International Institute for Population Sciences 2007, National Family Health Survey (NFHS-3), 2005-06, India. V. II. Mumbai: International Institute for Population Sciences; p. 168.

Pi -Sunyer, FX 1994, Obesity. Shils, ME, Olson, JA and Shike, M (eds) *Modern nutrition in health and disease*. pp. 984-1006. Lea and Febiger, London.

Roy Burman, JJ 2012,' Status of tribal women in India'. *Mainstream*, Vol. L, no.12.

Sobol, J, and Stunkard, AJ 1989, 'Socio-economic status and obesity: A review of the literature'. *Psychological Bulletin,* Vol. 105, pp. 260-275.

World Health Organization 2000, *Obesity, preventing and managing the global epidemic. Report of World Health Organization consultation,* technical report series, No. 894., World Health Organization, Geneva.

World Health Organization 2003, *Diet, nutrition and the prevention of chronic diseases*. Report of a joint WHO/FAO expert consultation. Technical report series no. 916. Geneva: WHO.

World Health Organization 2013, *Obesity and overweight fact sheet N°311. Washington DC. (*http://www.who.int/mediacentre/factsheets/fs311/en, accessed 3 December 2013)

Zahol, LV 2010, Status of tribal women (online). Available: <"http:// /1/status.pdf. (Oct. 2012)

Chapter - 2

Physical Growth and Nutritional Status among the Children of Northeast India

— Dali Dutta and Sarthak Sengupta

Introduction

Growth study of a population has recently assumed considerable significance in understanding the health and nutritional status of children. The multifarious importance of growth studies in national life can never be set aside. Such studies not only provide valuable information about the rates and patterns of physical growth of the children but also reveal certain important, fundamental and biological facts which reflect the image of health and nutritional status of a country or region in general and children of that region in particular. Growth study proves the nutritional and clinical status of a population as well as provides relevant information to the developmental programmes undertaken by the government and non-government organization towards the welfare of the same. The growth of children is affected by both genetical and environmental factors. Some measurements like height, weight, and girth provide valuable information regarding the growth trend and nutritional status of children belonging to different population groups. Physical growth of the children may help in the assessment of the nutritional status of the children. Nutrition influences growth and development before as well as after birth. Environmental influences, especially nutrition, are of greater importance than genetic background in children's growth and development.

There are different measures of nutritional status assessment, such as clinical examination to detect signs and symptoms of nutritional deficiencies like goiter to detect iodine deficiency, eye examination to detect Vitamin A deficiency; laboratory techniques to detect decreased levels or activities of nutrients in body tissues, or of enzymes. However, the majority of the laboratory methods are invasive and thus poorly suited to general use in projects. The anthropologists used the anthropometric measurements to assess nutritional status.

The present chapter reports the growth and nutritional status of children from different population groups of Northeast India with special emphasis on the Ahom population of Assam. Cross-sectional anthropometric measurements like height and weight and indices like weight-for-age, height-for-age and Body Mass Index-for-age of children from 6 to 20 years of age is taken into consideration. The physical growth patterns in respect to body weight, height, and nutritional status of Ahom children is compared with other population groups of Northeast India.

For comparison we have taken into consideration data reported from among the Rabha (Choudhury, 1979), Assamese Muslim (Begum, 1995), Bengali Muslim boys (Hussain, 2003), Sonowal Kachari (Singh, 2008), Mishing (Sikdar, 2010), Muslims of Assam (Medhi, 2013); Deori (Deori, 2016) of both Assam and Arunachal Pradesh; Meitei (Devi, 1986) of Manipur; Hajongs (Das, 1986), War Khasi (Khongsdier 1995), Pnar[1] (Dkhar, 1996), Khynriam Khasi (Mukherjee, 2002), Khasi (Wariri, 2009), Pnar[2] (Kamar, 2011) of Meghalaya; and Lotha Naga (Tsopoe, 2003), Ao (Longkumer, 2013) of Nagaland besides one more Ahom[1] sample reported by Hazarika (1978). The nutritional status of the present Ahom children is compared with Assamese Muslim (Begum, 1995), Sonowal Kachari (Singh, 2008), Mishing (Sikdar, 2010), Ao (Longkumer, 2013), Muslim of Assam (Medhi, 2013) and Deori (Deori, 2016) of Assam and Arunachal Pradesh.

The Ahom is considered as one of the numerically dominant populations of Assam having Mongoloid ethnic strain. They belong to Tai - Mao or the Mao section of the Tai that is widely known as Shan. The Ahom people first came to India from Myanmar in 1228 A.D. and was expanded their authority up to 1826 A.D. At present, Dhemaji, Dibrugarh, Lakhimpur, Sibsagar, Majuli, and Tinsukia districts of upper Assam are the main area of concentration of the Ahom people.

Material and Methods

In the present study, we are mainly concerned with the cross-sectional approach of growth study. Cross-sectional anthropometric data of 1741 children (842 boys and 899 girls) aged 6 to 20 years of Ahom of Upper Assam were collected. Ages were collected on the basis of school records and cross-checked by their parents. Two anthropometric measurements, i.e., height and weight are only considered in the present study. Height was measured with an anthropometric rod to the nearest 0.1 cm while the weight was recorded nearest to the 0.1 kg using a portable weighing machine with light apparel and without shoes following standard anthropometric technique as recommended by Weiner and Laurie (1981).

The physical growth patterns of the Ahom children with respect to body weight, height, and few nutritional anthropometric indices are compared with children of other population groups of Northeast India. The weight and height of present Ahom children are compared with the Ahom[1], Rabha, Assamese Muslim, Bengali Muslim boys, Sonowal Kachari, Mishing, and Muslims of Assam; Meitei of Manipur; Hajongs , Pnar[1], Pnar[2], Khasi, War Khasi, Khynriam Khasi of Meghalaya; Lotha and Ao Naga of Nagaland, and Deori of Assam and Arunachal Pradesh. The nutritional status of present Ahom children is compared with Assamese Muslim, Sonowal Kachari, Mishing, Muslim, Deori, and Ao Naga.

Results and Discussion

The comparison of the stature of Ahom boys with other population groups of Northeast India (Table 2.1a and 2.1b) shows that the Ahom boys are taller than the Assamese Muslim, Bengali Muslim, Muslim boys of both Char and Non-Char area, and Mishings of Assam, Khynriam Khasi, War Khasi, Pnar[1], Pnar[2] of both urban and rural area of Meghalaya; and Lotha and Ao Naga of Nagaland. The Rabha and Sonowal Kachari boys of Assam have higher mean value particularly at 11, 13 and 14 years of age in respect of stature. The Rabha and Deori boys of Assam and Arunachal Pradesh exceed in stature at the age group of 6 years. The Hajong boys of Meghalaya are taller than the Ahom boys in the age groups 6, 8 and 10 years of age. The present Ahom population is shorter than the Ahom sample reported earlier by Hazarika (1978) in the age groups 6 and 8 to14 years of age.

The comparison of the stature of Ahom girls with other population groups of Northeast India (Table 2.2a and 2.2b) shows that the Ahom girls are taller than the Rabha, Assamese Muslim, Muslim of both char and non-char areas of Assam; Meitei of Manipur; Hajong, Khynriam Khasi, War Khasi, Khasi, and Pnar[1] of Meghalaya; and Ao Naga of Nagaland. The only exception to it is seen among the Rabha and Assamese Muslim girls in the age group of 6 years exhibit higher stature than the Ahom girls. The Sonowal Kachari girls are taller than the Ahom girls at the ages of 10 to 15 and 18 to 20 years. The Mishing girls are taller than the Ahom girls at the adolescents' periods (10 - 14 years of age). The Deori girls of Assam are taller than the Ahom girls in the age groups 6,8,12, 13 and 18 years of age

Table 2.1a: Comparisons of the stature of Ahom boys with other population groups of Northeast India

Age (in years)	*Ahom[1]*	*Rabha*	*Meitei*	*Assamese Muslim*	*War Khasi*	*Khynriam Khasi*	*Bengali Muslim*	*Lotha Naga*	*Sonowal Kachari*	*Mishing*	*Pnar[2]*
6	111.39	110.15		107.14	107.98	103.21	106.81		106.66	105.63	
7	115.87	116.97		114.17	113.86	108.69	111.09		114.58	112.85	
8	126.57	120.08		120.56	118.96	113.61	117.00		119.93	117.36	114.49
9	135.39	125.98		124.58	124.20	119.20	121.90	119.49	126.12	122.83	120.39
10	141.03	129.86		130.50	127.52	122.76	125.93	124.07	131.32	127.74	123.26
11	143.11	138.06		135.69	132.44	126.13	131.32	130.60	137.77	135.69	129.85
12	147.63	141.36	136.40	141.16	137.75	129.22	133.51	134.84	143.28	141.57	135.04
13	154.37	149.63	139.70	144.20	142.85	136.94	139.59	145.06	150.49	145.66	140.71
14	160.33	154.54	146.45	154.05	149.87	140.64	149.12	151.77	160.19	148.51	148.08
15	161.73	158.88	155.70	159.30	153.11	144.25	155.51	156.43	162.23	149.35	154.05
16	163.67	161.33	159.20	161.13		146.78	159.80	160.24	162.94	150.58	157.05
17	163.77	161.33	163.05	162.11		150.27	160.93	162.37	163.41	151.59	159.04
18	164.11	161.41	164.25	163.28		154.65	162.79	162.79	164.06	162.55	160.30
19	165.13			164.71					164.93	163.51	
20	166.57			165.36					164.22	164.22	

Sources: Ahom[1] (Hazarika, 1978); Rabha (Choudhury, 1979); Meitei (Devi, 1986); Assamese Muslim (Begum, 1995); War Khasi (Khongsdier, 1995); Khynriam Khasi (Mukherjee, 2002), Bengali Muslim (Hussain, 2003); Lotha Naga (Tsopoe, 2003); Sonowal Kachari (Singh, 2008); Mishing (Sikdar, 2010); Pnar[2] (Kamar, 2011); Ahom (Present study).

Table 2.1b: Comparisons of the Stature of Ahom boys with other population groups of Northeast India

Age in years	*Hajong*	*Pnar[1]*		*Khasi*	*Ao*	*Muslim*		*Deori*		*Ahom (Present Study)*
		Urban	*Rural*			*Char area*	*Non-char area*	*Assam*	*Arunachal Pradesh*	
6.	110.82			106.06		103.5	104.5	110.85	112.75	109.68
7.	116.94			111.52		106.6	106.4	117.61	118.00	119.66
8.	125.78			116.90	122.49	107.8	107.8	120.63	122.22	123.26
9.	128.82			121.42	124.66	109.9	109.6	125.38	125.97	129.30
10.	132.51			125.62	128.36	112.1	115.2	130.90	129.85	131.49
11.	134.21	127.40	123.92	129.43	133.81	116.9	119.5	135.14	137.67	137.71
12.	137.06	131.94	130.55	135.38	139.41	122.2	123.0	141.43	143.00	144.14
13.	144.49	143.12	139.45`	143.26	147.86	125.9	129.3	146.94	148.94	148.73
14.	148.94	147.88	146.28	148.79	154.05	132.7	136.3	157.68	159.63	156.43
15.	155.32	155.40	152.58	152.71	161.15	137.0	137.3	160.96	161.01	163.65
16.	157.64	157.59	156.00	154.64	161.26	140.2	139.6	162.48	162.73	164.18
17.	161.89	158.33	156.38	155.60	162.67	143.1	143.8	163.41	164.05	165.97
18.	163.75	160.92	158.28	157.37	165.17	145.9	146.9	164.68	164.66	166.74
19.										166.97
20.										167.65

Sources: Hajong (Das, 1986); Pnar[2] (Dkhar, 1996); Khasi (Wariri, 2009); Ao (Longkumer, 2013); Muslim (Medhi, 2013); Deori (Deori, 2016).

Table 2.2a: Comparisons of the stature of Ahom girls with other population groups of Northeast India

Age (in years)	Rabha	Meitei	Assamese Muslim	War Khasi	Khynriam Khasi	Lotha Naga	Sonowal Kachari	Mishing	Pnar[1]
6	109.78	104.89	108.35	107.24	103.43		105.96	105.63	
7	115.17	108.61	114.74	111.23	108.65		113.78	112.85	
8	119.91	113.61	119.97	116.02	114.19		119.86	117.36	116.56
9	127.40	120.73	122.57	121.27	118.87	119.08	126.03	122.83	120.96
10	131.77	125.27	129.48	127.15	122.63	124.75	132.67	127.74	126.47
11	138.49	131.23	136.71	133.17	126.24	132.21	139.84	135.69	132.33
12	141.64	136.19	142.70	139.80	135.09	138.14	146.81	141.57	136.76
13	147.14	143.97	146.95	142.36	137.80	143.81	149.79	145.66	142.13
14	148.92	146.87	148.45	143.42	142.22	148.40	152.94	148.51	145.09
15	148.93	148.13	150.18	144.38	143.72	151.92	153.12	149.35	148.75
16	149.90	151.41	150.51		145.84	154.61	153.74	150.58	148.84
17	149.94	152.09	150.51		146.89	156.25	154.39	151.59	148.94
18	150.28	153.49	150.95		147.20	156.73	155.11	152.32	150.36
19			151.75				154.67	155.58	
20			151.81				154.67	156.55	

Sources: Rabha (Choudhury, 1979); Meitei (Devi, 1986); Assamese Muslim (Begum, 1995); War Khasi (Khongsdier, 1995); Khynriam Khasi (Mukherjee, 2002), Lotha Naga (Tsopoe, 2003); Sonowal Kachari (Singh, 2008); Mishing (Sikdar, 2010); Pnar[1] (Kamar, 2011); Ahom (Present study).

Table 2.2b: Comparisons of the Stature of Ahom girls with other population groups of Northeast India

Age in years	Hajong	Khasi	Ao	Muslim		Deori		Ahom (Present Study)
				Char area	Non char area	Assam	Arunachal Pradesh	
6	107.95	104.97		102.2	102.0	109.41	110.35	107.28
7	115.89	110.40		105.0	104.3	115.65	116.79	117.96
8	119.34	115.12	124.08	106.3	106.1	123.36	122.77	121.83
9	121.84	120.88	125.88	108.7	107.4	126.04	127.16	128.70
10	127.82	124.86	130.17	111.4	113.4	131.39	132.24	131.72
11	135.95	130.10	134.65	113.5	119.0	137.54	139.38	139.72
12	142.65	137.34	142.90	117.7	126.5	145.76	147.85	144.36
13	146.64	141.49	149.76	126.9	130.0	148.95	149.40	147.77
14	148.23	144.53	149.88	133.0	133.0	150.22	151.30	151.29
15	150.99	147.51	151.93	134.9	136.9	152.38	152.87	152.32
16		148.93	152.21	138.0	139.6	153.42	153.98	154.05
17		150.00	152.75	140.9	141.7	154.55	155.20	154.68
18		150.64	153.75	142.3	143.0	155.26	155.98	154.74
19								155.31
20								156.11

Sources: Hajong(Das, 1986); Khasi (Wariri, 2009); Ao (Longkumer, 2013); Muslim (Medhi, 2013); Deori (Deori, 2016).

while the Deori girls of Arunachal Pradesh are taller in the age group of 10 years. The Lotha Naga girls, on the other hand, are taller than the Ahom girls in between the ages of 16 to 18 years.

The comparison of the mean weight of Ahom boys with other population groups of Northeast India (Table 2.3a and 2.3b) shows that the Ahom boys are heavier than the Khynriam Khasi, War Khasi of Meghalaya, and Lotha Naga of Nagaland in all the age groups with the only exception at 6 years of age. The mean weight of Ahom boys when compared with Rabha and Sonowal Kachari of Assam show marked deviation in the age group of 6, 8, 11, 13 and 14 years, the Ahom boys are lighter than the Rabha boys while in the age groups of 10 to 14 years, the Sonowal Kachari boys surpass the Ahom boys. The Assamese Muslims, Bengali Muslims, and Muslim of Char and non-Char areas of Assam, Meitei of Manipur; both Pnar[1], Pnar[2] sample from Meghalaya; and Lotha Naga boys of Nagaland are lighter than the Ahom boys in all the ages. The Mishing boys of Assam are heavier than the Ahom boys in the age groups 10 and 11 years. Similarly, the Hajong and Khasi boys of Meghalaya are heavier in the age groups 6 and 8 years of age. The Ao boys of Nagaland are heavier than the Ahom boys in the 8 and 14 years of age. On the other hand, the Deori boys of both Assam and Arunachal Pradesh are heavier than the Ahom boys in all the age groups. One sample of Ahom boys studied earlier (Hazarika, 1978) show higher values at the post-adolescent periods i.e. 16- 20 years of age.

The mean body weight for the Ahom girls and other population groups of Northeast India (Table 2.4a and 2.4b) reports that in all the stages, the Ahom girls are heavier than the Khasi, Khynriam Khasi, War Khasi and Hajong of Meghalaya; Ao, and Lotha Naga of Nagaland in most of the age groups. On the other hand, the Rabha and Sonowal Kachari of Assam; and Meitei girls of Manipur have higher mean values in respect of weight at different ages. The Pnar[1] of Meghalaya and Deori girls of Assam and Arunachal Pradesh are heavier than the Ahom girls in all the age groups. Muslim girls of Char and Non-Char of Assam are lighter than the Ahom girls in all the age groups. On the other hand, the Mishing and Assamese Muslims girls are lighter than the Ahom girls but the exception is seen in the age groups 6 years of age.

The incidence of underweight in Ahom boys (23.99%) is slightly higher than the Lotha Naga (22.41%), and Sonowal Kachari boys (23.12%). On the other hand, the frequency of underweight is quite frequent among the Mishing (30.10%), Assamese Muslim (46.95%), Muslim (51.45%) of Assam. The incidence of underweight is quite low among the Deori (7.94%)

Table 2.3a: Comparisons of the weight of Ahom boys with other population groups of Northeast India

Age (in years)	*Ahom*[1]	*Rabha*	*Meitei*	*Assamese Muslim*	*War Khasi*	*Khynriam Khasi*	*Lotha Naga*	*Bengali Muslim*	*Sonowal Kachari*	*Mishing*	*Pnar*[1]
6	17.69	17.93		16.69	17.83	17.10		15.17	16.17	16.50	
7	18.52	20.17		19.33	19.69	18.11		16.55	18.71	19.11	
8	23.67	21.59		20.90	21.25	20.65		18.01	20.67	21.34	19.19
9	27.71	23.96		22.20	23.48	23.04	22.50	20.12	23.90	23.72	21.63
10	30.51	25.70		26.02	25.40	24.59	22.92	21.25	26.23	26.47	22.48
11	32.55	29.86		27.52	27.71	26.98	29.15	22.96	29.94	29.38	26.18
12	36.23	32.02	29.27	31.52	30.36	28.86	31.82	25.33	33.44	32.47	29.00
13	41.71	38.61	32.52	34.66	33.67	34.00	35.97	28.32	38.89	35.26	33.01
14	47.99	41.82	37.49	39.71	37.26	37.30	40.23	34.19	45.83	40.04	38.07
15	48.27	45.40	44.46	45.23	40.49	41.67	43.86	39.66	48.53	43.38	43.04
16	52.11	49.16	47.05	46.75		43.29	47.51	43.76	50.35	46.13	47.77
17	52.57	51.04	50.75	49.58		47.63	49.31	44.88	51.17	48.09	51.02
18	54.91	51.08	51.96	49.70		48.88	49.81	46.22	53.13	50.29	51.21
19	56.05			51.26					54.32	51.27	
20	55.81			51.56					54.98	52.35	

Sources: As shown in Table 1a.

Table 2.3b: Comparisons of the weight of Ahom boys with other population groups of Northeast India

Age in years	***Hajong***	***Pnar***[2]		***Khasi***	***Ao***	***Muslim***		***Deori***		***Ahom (Present Study)***
		Urban	***Rural***			***Char area***	***Non-char area***	***Assam***	***Arunachal Pradesh***	
6	17.32			18.79		15.5	16.2	20.09	20.97	17.05
7	18.72			19.54		16.3	17.5	22.71	23.01	20.33
8	22.27			22.17	22.94	17.8	18.5	23.61	24.04	21.42
9	23.00			21.87	24.06	20.0	19.4	25.76	26.63	24.73
10	23.25			25.17	25.22	24.0	23.1	28.02	27.64	25.74
11	24.85	24.38	23.78	27.30	28.42	25.2	27.5	31.31	31.74	28.51
12	29.37	28.87	27.13	30.18	31.79	28.0	29.7	35.34	35.24	32.63
13	35.42	35.03	34.13	38.77	36.64	32.4	34.7	40.79	39.91	36.18
14	38.83	41.07	38.86	44.98	42.18	38.8	38.0	47.33	48.84	41.78
15	41.50	46.36	44.16	46.32	48.77	42.0	40.7	51.33	51.86	49.20
16	45.15	49.33	47.16	48.77	49.03	44.7	43.2	52.40	54.49	50.11
17	47.90	50.40	49.55	50.90	50.91	48.3	46.7	54.78	57.04	55.08
18	48.12	54.21	52.03	52.30	54.07	50.6	49.7	57.82	58.69	55.67
19										58.02
20										59.60

Sources: As shown in Table 1b

Table 2.4a: Comparisons of the weight of Ahom girls with other population groups of Northeast India

Age (in years)	*Rabha*	*Meitei*	*Assamese Muslims*	*War Khasi*	*Khynriam Khasi*	*Lotha Naga*	*Sonowal Kachari*	*Pnar*[1]	*Mishing*
6	16.81	17.89	17.21	17.00	17.74		15.51	18.94	16.37
7	19.07	19.65	19.02	18.82	18.69		18.73	21.44	18.83
8	20.97	22.75	20.90	20.27	20.41		20.19	24.49	20.28
9	24.50	24.41	22.05	22.65	22.68	22.50	23.73	28.35	22.45
10	26.51	25.73	25.18	24.74	24.72	22.92	26.25	30.48	24.77
11	30.69	26.81	29.63	27.98	26.99	29.15	31.74	36.62	27.71
12	34.66	30.09	33.68	31.58	30.87	31.82	38.29	39.92	32.68
13	39.69	33.83	38.70	33.54	33.08	35.97	42.42	43.83	36.50
14	42.81	37.33	41.05	35.38	37.27	40.23	44.23	44.89	39.70
15	43.05	42.29	42.52	37.22	41.12	43.86	45.40	45.05	41.34
16	46.90	44.45	42.71		43.17	47.51	46.88	47.66	42.42
17	46.96	46.75	43.43		44.67	49.31	48.34		42.99
18	47.19	49.01	43.51		44.38	49.81	49.82		44.28
19		49.03	44.76				51.27		44.80
20		49.05	45.44				52.48		45.50

Sources: As shown in Table 2a

Table 2.4b: Comparisons of the weight of Ahom girls with other population groups of Northeast India

Age in years	*Hajong*	*Khasi*	*Ao*	*Muslim*		*Deori*		*Ahom (Present Study)*
				Char area	*Non-Char area*	*Assam*	*Arunachal Pradesh*	
6	16.22	16.97		15.2	15.4	19.62	20.32	15.91
7	18.27	19.73		17.0	16.4	21.45	22.32	19.98
8	19.78	21.18	22.86	18.0	17.8	24.83	25.67	20.53
9	22.79	22.82	23.77	20.1	18.5	26.15	26.45	24.20
10	26.22	25.63	25.82	23.2	21.9	28.48	29.88	25.93
11	28.60	27.57	29.22	26.0	23.8	32.42	33.33	31.58
12	33.00	32.25	34.82	28.1	29.9	38.41	40.02	34.66
13	36.00	35.92	41.22	32.1	30.0	43.02	44.77	38.92
14	37.45	42.31	43.82	35.9	33.0	45.62	46.99	42.49
15	41.32	44.33	46.75	38.1	35.7	48.89	48.74	44.27
16		46.10	45.65	39.9	37.9	49.43	50.19	45.98
17		47.77	46.63	43.1	40.0	50.23	51.53	46.02
18		47.07	48.51	45.8	42.3	51.13	52.28	46.84
19								47.12
20								48.85

Sources: As shown in Table 2b.

of Assam and Arunachal Pradesh. The Ahom girls are comparatively more underweight (21.91%) than the Lotha Naga (16.53%), Deori (2.51%) and Sonowal Kachari (16.13%) girls. However, the incidence of underweight is quite common among the Mishing (33.30%), Assamese Muslim (47.74%) and Muslim (52.87%) of Assam.

The frequency of stunted Ahom boys (15.20%) is more than the Lotha Naga (13.31%) boys of Nagaland. On the other hand, the frequency is quite common among the Sonowal Kachari (17.41%), Deori (28.26%), Mishing (28.90%), Assamese Muslim (20.56%), Muslim (41.48%) of Assam and Khynriam Khasi (95.71%) of Meghalaya. The Ahom girls (17.69%) are comparatively less stunted than the Sonowal Kachari (23.32%), Deori (28.25%), Mishing (33.30%), Assamese Muslim (21.11%), Muslim (43.95%) of Assam, Khynriam Khasi (94.96%) of Meghalaya and Lotha Naga (24.53%) of Nagaland.

With respect to the prevalence of chronic energy deficiency cases, it is seen that the Ahom boys are more in number in the wasted category (22.09%) than the Mishing (20.70%) and Sonowal Kachari (10.13%) and Deori (8.67%), boys of Assam; Khynriam Khasi (14.29%) of Meghalaya, and Lotha Naga (8.20%) of Nagaland, However, the incidence of wasting is frequently met with among the Assamese Muslim (65.48%) and Muslim (51.12%) of Assam; and Ao Naga (31.44%) of Nagaland. In the present study, 12.57% of Ahom girls are recorded in the wasted category and such incidence is relatively low in Lotha Naga (5.07%), Sonowal Kachari

Table 2.5: Comparison of Nutritional Status of Ahom with other population groups of Northeast India

Population	*Underweight (%)*		*Stunted (%)*		*Low BMI for age (%)*		*Source*
	Boys	*Girls*	*Boys*	*Girls*	*Boys*	*Girls*	
Ahom	23.99	21.91	15.20	17.69	22.09	12.57	Present Study
Assamese Muslim	46.95	47.74	20.56	21.11	65.48	68.59	Begum, 1995
Sonowal Kachari	23.12	16.13	17.41	23.32	10.13	6.10	Singh, 2008
Mishing	30.10	33.30	28.90	33.30	20.70	18.00	Sikdar, 2010
Ao					31.44	25.13	Longkumer, 2013
Muslim (Char & Non Char)	51.45	52.87	41.48	43.95	51.12	53.50	Medhi,2013
Deori	7.94	2.51	28.26	28.25	8.67	6.81	Deori, 2016

(6.06%), and Deori (6.81%) as very few girls represent the wasted category. On the other hand, among the Khynriam Khasi (13.38%), Mishing (18.00%), Ao (25.13), Assamese Muslim (68.59%) and Muslim (53.50%) incidence of wasting is comparatively more than the Ahom girls.

The Ahom children are taller and heavier than the Assamese Muslims, Bengali Muslims, Muslim of Assam, and Meitei of Manipur. They are also taller than Khynriam Khasi, War Khasi, Pnar of Meghalaya and heavier than the Lotha Naga, and Pnar of Meghalaya in all the age groups. On the contrary, the Deori children are heavier than the Ahom children in all the age groups.

The nutritional status according to weight for age, height for age and chronic energy deficiency, the Ahom children stands relatively in a better position than the children of Assamese Muslim and Mishing of Assam. Such a better growth and nutritional status reflect the better dietary intakes of the Ahom population. The present study on children of Northeast India in general revealed poor nutritional state. Children of all the population groups certainly need appropriate support in the domains of health and nutritional programmes in an effective manner so that their health and nutritional status can be better managed.

Reference

Begum, G 1995, *Age changes in some somatometric characters with reference to the nutritional status of Assamese Muslims of Kamrup district, Assam*. Ph.D. thesis, Gauhati University.

Choudhury, B 1979, *A cross-sectional study of the growth of some somatometric characters of Rabha boys of Kamrup district, Assam.* Ph.D. thesis, Gauhati University.

Das, H 1986, *Age changes in some somatometric characters of the Hajongs of Meghalaya.* Ph.D. thesis, Gauhati University.

Deori, J 2016, *A cross-sectional study on physical growth among the Deoris of Assam and Arunachal Pradesh*. Ph.D. thesis, Gauhati University.

Devi, S 1986, 'A comparative study of growth with regard to height and weight between the Meitei and Muslim girls of Manipur'. *Bulletin of the Department of Anthropology, Dibrugarh University,* Vol. 15, pp. 140-150.

Dkhar, JW 1996, *Age changes in somatotypes of 11 to 18-year-old Pnar boys of Meghalaya.* Ph. D thesis, North Eastern Hill University.

Hazarika, ANZA 1978, *A study of the growth of head, face, trunk, and limbs of the Ahom boys of Assam.* Ph.D. thesis, Dibrugarh University.

Hussain, A 2003, *Somatotypes, physical growth and nutritional status of the Bengali Muslim boys of Barak Valley, Assam: A bio-anthropological Study.* Ph.D. thesis, Gauhati University.

Kamar, P 2011, *Physical growth and nutritional status among the Pnar adolescents of Jaintia Hills, Meghalaya*. Ph.D. thesis, Northeastern Hill University.

Khongsdier, R 1995, *A study on nutrition and health status in relation to some biosocial factors among the War Khasi of Meghalaya*. Ph.D. thesis, North Eastern Hill University.

Longkumer, T 2013, *A study on somatotype and nutritional status in relation to physical activities among the Ao children of Mokokchung district Nagaland.* Ph.D. thesis, Northeastern Hill University.

Medhi, D 2013, *Growth and nutritional status among the Muslims of Barpeta district Assam.* Ph. D thesis, Gauhati University.

Mukherjee, N 2002, *Studies on demography and growth pattern among the Khasi children of Shillong, Meghalaya.* Ph.D. thesis, Northeastern Hill University.

Sikdar, M 2010, *Biosocial determinants of growth and nutritional status of the Mishings of upper Assam*. Ph. D thesis, Dibrugarh University.

Singh, J 2008, *A study on growth and nutritional status of the Sonowal Kachari of Dibrugarh district, Assam*. Ph.D. thesis, Dibrugarh University, Dibrugarh.

Tsopoe, T 2003, *Nutritional status, and physical growth of Lotha children of Wokha district, Nagaland.* Ph.D. thesis, North Eastern Hill University.

Wariri, BC 2009, *Growth and nutritional status of the Khasi children in West Khasi Hills district of Meghalaya*. Ph.D. thesis, North Eastern Hill University.

Weiner, JS and Lourie, JA 1981, *Practical human biology*. Academic Press, London.

Chapter - 3

Age-Related Changes in the Anthropometric and Body Composition Characteristics of the Garo Women of Kamrup District, Assam

— *Gulrukh Begum and Flora Rabha*

Introduction

Age-related changes describe changes that occur with advancing age. Normally physiological changes attain a maximum level between the late teens and thirty years of age. After 35 years there occurs a decline in physiological performance measures. Different characters change at different rates in different people. Chronological age does not necessarily correlate with biological age (Jylhava, 2017). There is great heterogeneity in the health outcomes of older individuals (Lowsky et al., 2014).

Of the numerous physiological changes that accompany advancing age, changes in body composition may be among the most significant in terms of health. An accurate assessment of body composition is necessary to correctly identify the health risks associated with excessively low or high body fat. Body weight reflects the proportion of fat and lean mass (bones, tissues, organs, and muscles). Considering the profoundness of the health problems associated with too much or too little body fat, it is no surprise that the assessment of body composition has become widespread and meaningful. Excessive body fat has been demonstrated to be associated with such health-related problems as coronary artery disease, hypertension, adult-onset diabetes, hyperlipidemia, obstructive pulmonary disease, and osteoarthritis. Too little fatness, as seen in individuals with eating disorders, exercise addiction, and certain diseases, such as cystic fibrosis, can also lead to serious physiological dysfunction (Heyward and Stolarczyk, 1996).

Body composition is the components that make up the body such as bone, muscle, fluid (high amounts are desirable), and fat (high amounts are undesirable). By taking both desirable and undesirable components of the body into account a better assessment of obesity and its associated health implications is done. There are plethoras of ways body composition can be assessed. The most popular field techniques include bioelectrical impedance analysis (BIA), skinfolds (SF) and anthropometry (Wagner and Heyward, 1999).

Anthropometric measurement data could be used to evaluate body composition changes that occur over the adult lifespan, health and dietary status, and disease risk (CDC, 2007). According to WHO (1995), anthropometry is a single most universally acceptable inexpensive and non-invasive method available to access the size, proportion, and composition of the human body. Anthropometry still is the most widely used method, and recently, it has been used to estimate fat distribution.

In adults, intra-abdominal adipose tissue (body fat around the visceral organs) is related to negative health outcome independent of total body fat. Thus, assessment of body fat distribution is equally as important as the measurement of total body fat. Traditionally skinfold thickness measurements are used for assessing relative fatness and to assess the size of specific subcutaneous fat deposits. Research on the distribution of body fat has shown that for both adults and children, a more central distribution of fat is associated with type 2 (non-insulin-dependent) diabetes mellitus, a worse cardiovascular disease risk profile, and other adverse outcomes (Kissebah et al.,1982; Thompson et al.,1991; Haffner et al.,1989).

The most common assessment of obesity currently is the Body Mass Index (BMI). BMI is inexpensive, non-invasive, and requires little training but is not a direct measure of body fat percentage (BF %), which has led to under-and over-estimation of adiposity based on ethnicity.

Athletes often have increased weight due to a high percent of muscle mass, which weights more than fat. Hence, these individuals have a low body fat percentage. However, when BMI is calculated with only height and weight, this population looks to be overweight or obese on the BMI scale (Ode et al., 2007). The reason for this disparity is that BMI does not take actual body composition into account. There are other anthropometric measures such as waist circumference (WC) and sagittal abdominal diameter (SAD) for estimating the body composition. Waist circumference (WC), presents high sensitivity and specificity to central fat evaluation in children and adolescents. Similarly to BMI, WC provides only a general overview of body composition.

Estimation of fat mass from anthropometry uses the anthropometric measures (e.g., the sum of skinfolds) which are related to body fat mass. A skinfold thickness measurement provides an estimated size of the subcutaneous fat deposit, which is basically the fat under the skin. By estimating the thickness of this area estimation of the total body fat could be done. The goal of this method is to provide an assessment of adiposity, which is a better method than the Body Mass Index (BMI). The measurement of skin folds is by nature sensitive to inter-user variability. Second, because of potential gender, ethnic, and maturation-related changes in body composition, the relationship between skinfold measures and body fat may vary between subgroups of the population. Thus, it is important to identify anthropometric measures that are robust to inter-user variability and equally reflective of body composition in all subgroups of the population (Goran, 1998).

Apart from BMI and circumferential measurements, alternative measurements of body composition, such as bioelectrical impedance analysis (BIA) are used and population-specific models have been created to use an individual's impedance value to estimate percentage body fat (%BF) (Kyle et al.,2004; Ricciardi and Talbot, 2007). Bioelectrical resistance is an alternative technique for assessing body composition in clinical and population-based studies. The technique is based on measurement of electrical resistance in the body to a tiny imperceptible current. The electrical resistance is a function of body shape and the volume of conductive tissues in the body (NIH, 1995). The measurement of the total body fat provides useful information. It may influence morbidity and mortality, it may affect the ability to withstand exposure to cold and starvation, and it may alter the effectiveness of drugs and anaesthetics (Durnin and Womersley, 1974). It may also be the cause of reduced insulin sensitivity in some population groups (Osei et al., 1991). Because of the health effects of excess body fat (e.g., increased risk of Type 2 diabetes, heart disease, cancer) and of being underweight (e.g., anorexia nervosa), and because of the deleterious effects of excess fat on athletic performance, quantifying body composition in terms of percent fat has important uses (e.g., for implementing and monitoring weight loss programs). In addition, quantifying fat-free mass and its components (muscle, bone) is important because of the health implications of sarcopenia and osteopenia/osteoporosis.

Body composition assessment is also done with the help of upper arm anthropometry. Upper arm anthropometry is composed of upper arm muscle area (UMA), total upper arm area (TUA), upper arm fat area (UFA), arm fat index (AFI), upper arm fat area estimate (UFE) and upper arm muscle area estimate (UME) is used to assess body composition. Upper arm muscle area (UMA), upper arm fat area (UFA) and upper arm muscle area by height (UAMAH) are also used in the assessment of nutritional status of children (Bolzan et al., 1999, Frisancho and Tracer, 1987). Body composition analysis is essential in assessing the nutritional status, as well as the risk of developing diseases associated with abnormal body fat content.

The anthropometric and BIA methods provide different results of body fat content, especially in women, so they should not be used interchangeably (Keska, 2012). The assessment of the body composition is an important measure of the nutritional status in man, because body fat (BF) is directly related to obesity and diet-related diseases, whereas low levels of fat-free mass (FFM) may be more critical to the health of infants and children, elderly, malnourished persons, maturating women, and those with muscle-wasting diseases (Bujko, 2006).

The variation in the anthropometric and body composition characteristics in adulthood is primarily due to osteological changes, changes in the fat and muscle tissue. Because of the cross-sectional nature of most of the anthropological studies on adults, it is difficult to separate whether the changes occur due to secular changes or from continuing growth or due to aging (Roy and Pal, 2003). The present study, therefore, is an attempt to provide the nature and extent of age-related changes in body composition characteristics using anthropometric analysis and BIA methods among the adult plains Garo women of Kamrup district, Assam.

Materials and Method

The present cross-sectional study was undertaken among the Plains Garo women of Borbaka and Manik Nagar villages which are about 25-30 km away from Guwahati city which falls under Kamrup district of Assam. The Garos are mainly distributed over the Garo Hills in Meghalaya, Kamrup, Goalpara, Karbi Anglong districts of Assam, Khasi Hills in Meghalaya and Dimapur (Nagaland State). Substantial numbers are found in districts of Bangladesh also. The highest concentration of the Garos is seen in the Garo hills district of Meghalaya followed by Goalpara, Karbi Anglong and Kamrup District of Assam.

The Garos are one of the Scheduled Tribes in the North-Eastern state of Meghalaya, India. The Garos call themselves Achik-Mande, which literally mean 'hill people', from a chik "bite soil" + mande "people" or simply A.chik or Mande (www.wikipedia.org). They are one of the few remaining matrilineal societies in the world. They are a matrilineal society but not matriarchal. The acceptance and conversion to Christianity have ushered in innumerable changes in the life and culture of the Garos through the efforts of the Christian Missionaries.

The study was carried out among 283 adult Plains Garo females from 20 to 60 yrs of age. Anthropometric data were collected from women who voluntarily agreed to participate in the study. Age of the individuals have been collected after verifying the written record, those individuals who did not have a birth record, their age was estimated by referring to some important local events. Total 283 females have been classified into 8 groups of 5 years age cohort, i.e., 20-25, 25-30, 30-35, 35-40, 40-45, 45-50, 50-55, 55-60 years. Pregnant and sick women were not included in the study.

Data types include socio-demographic characteristics of the individuals studied like age, sex, education, occupation as well as their anthropometric measurements. Anthropometric data have been collected following standard techniques. The measurements include height, weight, mid-upper arm circumference, waist circumference, hip circumference, and skinfold thickness at biceps, triceps, subscapular, suprailiac and medial calf. These were recorded using standard anthropometric procedures (Gibson, 2005). Heights of the subjects were recorded with the help of an anthropometer with the head of the subject being held in the Frankfurt Horizontal plane to the nearest 0.10 cm. The weight was recorded using a portable digital weighing machine to the nearest 100g. The MUAC was taken at the point midway between the acromion and the radiale of the upper-arm using a plastic coated non-stretchable measuring tape. Waist circumference was measured at the minimum circumference between the iliac crest and the rib cage. Hip circumference was measured at the maximum protuberance of the buttocks.

Skinfold at the following sites was measured using a Harpenden skinfold caliper calibrated to exert a constant pressure of $10 gm/mm^2$ on the left side to the nearest 0.20 mm:

1. Triceps, halfway between the acromion process and the olecranon process.
2. Biceps, at the same level as the triceps skinfold, directly above the center of the cubital fossa.
3. Subscapular, about 20 mm below the tip of the scapula, at an angle of 45° to the lateral side of the body.
4. Supra-iliac, about 20 mm above the iliac crest, in the axillary line.

Assessment of upper arm composition was evaluated using standard equations based on MUAC and TSF utilizing the standard equations of Frisancho (1981, 1989):

$$TUA\ cm^2 = (MUAC)^{2/}(4\times ð)$$
$$UMA\ cm^2 = \{MUAC - (TSF\times ð)\}^{2}/(4\times n)$$
$$UFA\ cm^2 = \{(MUAC)^2/(4\times ð)\} - UMA$$

Body composition was evaluated using Omron Body Fat Analyzer, which works on the principles of Bioelectrical Impedance to measure a person's actual body fat content. By utilizing BIA (Bioelectrical Impedance Analysis) principles, a relatively accurate picture can be obtained of the body's fat content. The method Bioelectrical Impedance is considered to be one of the standard techniques (Kushner et al., 1990).

Fat mass (FM) and Fat-free mass (FFM) data have been calculated by using the following formula:

FM (kg) (Fat mass) = (Body Fat percent/100) × Weight (kg)
FFM (kg) (Fat free mass) = Weight (kg) – Fat mass (kg)

Some anthropometric indices were also calculated for the body composition analysis:

Centripetal fat ratio, Subscapular Triceps Skin fold Ratio (SSTSFR), Subscapular Supra iliac Triceps-Bicep Skin fold Ratio(SSI-TBSFR), Sum of Truncal Skin fold (STSF), Sum of Peripheral Skin fold (SPSF), Sum of Truncal and Peripheral Skin fold Ratio (STPSFR).

Analysis of data

All descriptive data pertaining to anthropometric measurements and body composition variables has been reported as mean and standard deviation. The data were analyzed using the Statistical Package for Social Science (version 17.0).

Results and Discussion

At a certain age, individuals reach physical maturity through growth and development but the age of maturity of all the anthropometric characters and body composition characters are not similar between individuals or populations. The present study examines age-related changes in anthropometric and body composition characteristics of the Garo women of Kamrup district Assam. The study population belongs to a similar socio-economic background and shares a common physical environment. Study protocols were similar for all the individuals studied.

Table 3.1: Age-specific descriptive statistics of anthropometric variables among the Garo women of Kamrup district, Assam

Anthropometric variables	*20 – 25 years N = 44*	*25 – 30 years N=39*	*30 – 35 years N=41*	*35 – 40 years N=40*	*40 – 45 years N=40*	*45 – 50 years N= 34*	*50 – 55 years N= 25*	*55 – 60 years N= 20*
	Mean ± SD	***Mean ± SD***	***Mean ± SD***	***Mean ± SD***	***Mean ± SD***	***Mean ± SD***	***Mean ± SD***	***Mean ± SD***
Height	150.56 ± 4.82	152.64 ±11.09	150.60 ± 4.25	150.76 ± 4.96	152.65 ± 9.29	153.45 ± 4.79	151.85 ± 3.98	150.67 ± 3.67
Body Weight	46.28 ± 7.91	50.02 ± 11.08	52.15 ± 13.85	54.04 ± 10.08	54.26 ± 10.17	55.75 ± 7.33	53.75 ± 8.85	53.28 ± 12.19
Waist circumference	71.4 ± 6.24	73.56 ± 11.34	77.36 ± 12.01	75.95 ± 7.95	75.89 ± 8.22	78.03 ± 8.98	78.37 ± 7.91	75.64 ± 10.37
Hip circumference	82.35 ± 7.12	88.00 ± 9.08	88.37 ± 10.83	87.88 ± 7.63	89.22 ± 9.68	90.62 ± 7.07	91.5 ± 8.95	90.29 ± 11.45
BMI	20.39 ± 3.06	21.42 ± 4.37	23.02 ± 6.25	26.62 ± 3.66	20.42 ± 4.04	22.97 ± 3.31	22.04 ± 4.50	23.40 ± 4.94
Body Density	1.104 ± 0.01	1.101 ± 0.01	1.10 ± 0.01	1.100 ± 0.009	1.101 ± 0.01	1.09 ± 0.009	1.10 ± 0.01	1.09 ± 0.01
Fat Mass	10.44 ± 4.86	13.27 ± 6.54	15.03 ± 8.55	15.35 ± 5.84	15.92 ± 5.79	16.25 ± 4.72	16.72 ± 6.47	16.50 ± 8.18
Fat Free Mass	46.06 ± 7.85	50.97 ± 10.22	52.67 ± 12.97	50.47 ± 9.09	48.02 ± 10.12	53.46 ± 7.28	50.48 ± 8.78	52.99 ± 12.11

Table 3.1 shows descriptive statistics pertaining to an anthropometric characteristic of 8 groups of 5 years age cohorts of Garo women from 20 years onwards. The present study shows maximum mean height in the age group of 45-50 years and thereafter it is slightly decreasing with the increase in age. This anthropometric character shows a decreasing trend with the increment of age after 50 years may be due to the shrinkage of inter-vertebral space and increased curvature of the spine. The mean weight of women seems to increase with age, the highest being in the age group (45-50 years) and the second highest being in the age group (35-40 years). Weight gain often occurs during the gaining process for various reasons, such as reduced physical activity. This is sometimes paired with decreased height as spinal issues become more pronounced with age (Lexell et al., 1988). Women usually gain weight until the age of 65years, and then begin to lose weight. Weight loss later in life occurs partly because fat replaces lean muscle tissue, and fat weights less than muscle. The age-related increase in fat and decline in muscle often occur simultaneously (Wannamethee et al., 2005).

Weight is not a suitable measure for assessing ideal body composition related to fitness because an increase in weight is due to an increase in fat-free mass (FFM) can be misinterpreted as an increase in body fatness. Waist circumference also shows the highest measure in the age cohort of 50-55 years. There is a continuous increase as we move ahead in the age cohorts except between 35-45 years when there is a decrease. Hip circumference also increases as we move from one age cohort to another, but the increase is not much and the maximum circumference is found between 50-55 years of age. The mean BMI of the Garo women in all the age cohorts falls under the normal category. Fat mass increases with age and in the fat-free mass the increase takes place till 35 years of age then afterward no regular trend could be noticed. Ageing process leads to muscle loss. Muscle strength declines from people aged less than 40 years to those greater than 40 years (Engelhardt and Kellar, 2014) and degeneration of muscle and accumulation of fat mass occurs during the older age (Das and Roy, 2010).

The body density is uniform throughout the ages. Muscularity and adiposity could be attributed to several associated factors like ethnicity, dietary intake, food habits, physical exercise and socio-economic status (Thibault and Pichard, 2012).

Obesity is generally classified as generalized obesity (GO) and abdominal obesity (AO). Individuals with obesity have higher rates of mortality and morbidity compared to nonobese individuals (WHO, 2009). Although BMI is commonly employed as the measure of general or overall adiposity in most observational studies, growing evidence suggests that a central (abdominal) fat distribution pattern, as reflected by a higher waist circumference (WC) or waist-to-hip ratio (WHR), may be a better measure of risk than elevated body weight particularly among older adults (Huang et al., 1999; Huxley et al., 2010; Visscher et al., 2001). Percentage of occurrence of obesity among the Garo women has been calculated.

Table 3.2 Women at health risk as per Waist-Hip Ratio

Range	*Health Risk Level*	*No. of Women*	*Percentage*
0.80 or below	Low risk	44	15.55
0.81 to 0.85	Moderate risk	97	34.28
0.86 or higher	High risk	142	50.18
Total		283	100

Table 3.2 shows the health risk level of Garo women on the basis of waist to hip ratio.WHR is used as a measurement of obesity, which in turn is a possible indicator of other more serious health conditions. In the present study 50.18 percent of women were found to fall in the high health risk level i.e. they have abdominal obesity. 34.28 percent of women were found

to fall in moderate health risk level and 15.55 percent of women were found to fall in the low health risk level.

The WHO states that abdominal obesity is defined as a waist-hip ratio above 0.90 for males and above 0.85 for females (WHO, 2008). It determines how much fat is stored on the waist, hips, and buttocks. Not all excess weight is the same when it comes to health risks. People who carry more weight around their mid-section (an apple-shaped body) are at higher risk for heart disease, type 2 diabetes and premature death than those who carry more of their weight in their hips and thighs (a pear-shaped body). Even if your BMI is within a normal range, your risk for the disease may be increased.

Table 3.3 Abdominal Obesity among the Garo women

Age Group (Years)	*n*	*Abdominal Obesity*	*Percentage (%)*
20-30	83	40	48.78
30-40	81	39	48.15
40-50	74	37	50.00
50-60	45	26	57.78

In Table: 3.3, cohorts of women with abdominal obesity or with waist to hip ratio above 0.85 are presented. Here the last age group i.e. age group 50-60 shows the maximum percentage (57.78%) of women with abdominal obesity. This might be due to the trend of accumulation of fat as individual ages. Abdominal obesity assessed by waist circumference or WHR predicts the risk of diabetes and CHD (Coronary Heart Disease) independent of BMI (Liu et al., 2010; Sehested et al., 2010). As with BMI, waist circumference or WHR appears to be a stronger predictor of diabetes in women. Due to the increased risk of morbidity and mortality, obesity is now being recognized as a disease in its own right.

In countries like India, the rise in obesity prevalence could be attributed to the increasing urbanization, use of mechanized transport, increasing availability of processed and fast foods, increased television viewing, adoption of less physically active lifestyles and consumption of more "energy-dense, nutrient-poor" diets (Bell et al., 2002; Misra et al., 2010). In the present population, which is a semi-urban population living in the fringes of Guwahati city have a lifestyle quite similar to the urban areas. They are not employed in full-time cultivation. Most of the men are also engaged in other activities like wage labourers and some hold small-scale jobs in the city. Women are confined to the household works. Women are also engaged in cultivation but as age increases, they are more limited to their houses. This could be the reason behind the higher percentage of WHR of 50.18 percent which increase with age. The increase in the prevalence of obesity among rural population may be due to rapid changes in the lifestyle in rural areas (Pradeepa et al., 2015).

Asian Indians have a greater predisposition to abdominal obesity and accumulation of visceral fat and this has been termed as "Asian Indian Phenotype" (Joshi, 2003; Deepa et al., 2006).

The present study population shows a higher prevalence of isolated abdominal obesity. It is also to be noted that in the present population alcohol consumption is also prevalent and its intake is culturally accepted among women as well. Alcohol use is one of the NCD (Non-Communicable Diseases) risk factors in this population.

Bioelectrical impedance analysis (BIA) is a safe, non-invasive and proper way to determine the body fat and lean mass which is easily applied. The individual's age, sex, ethnic characteristics, hydration status, some medical conditions, environmental conditions, physical activity, personal characteristics are important factors that affect the results of BIA. It provides an estimation of the contribution of both lean and fat mass to body weight via an estimate of total body water.

Table 3.4 Obesity type among the Garo women

Obesity type	*No. of Women*	*Percentage*
Obese	65	22.97
Latent obesity	32	11.31
Muscular	4	1.41
Normal	114	40.28
Lean	68	24.03
Total	283	100.00

Table 3.4 shows percentage of women as per obesity type evaluated with BIA (Bioelectrical impedance analysis). Among the Garo women, 40.28 percent are normal, 24.03 percent are lean, 22.97 percent of women are obese, 11.31 percent of women have latent obesity, and 1.41 percent women are muscular. Latent obesity (L-obesity) is defined as BMI<25 with a body fat percentage (% fat) $\geq$ 30.00 percent. Latent obesity occurs when someone with a normal or low BMI has a higher than normal body fat percentage. It's sometimes called being "skinny fat" and it is a risk factor for the development of diabetes.

Table 3.5: Obesity among Garo women at 10 years age cohort

Age Group (in yrs)	*No. of Women*	*No. of Obese Women*	*Percentage (%) of Obese Women*
20-30	83	7	8.43
30-40	81	24	29.63
40-50	74	21	28.38
50-60	45	13	28.89

Table 3.5 show age groups with 10 year age cohorts of women with obesity. After 30 years of age, the percentage of occurrence of obesity has increased. This increase is almost similar in the subsequent age cohorts which are less than 30 percent. The prevalence of obesity has been found to be increasing with age and is higher in each younger generation (unfavourable generation shift) [Hulsegge et al., 2016]. This may be due to degeneration of muscle tissue and accumulation of fat during old age.

Subcutaneous adipose tissue measurements are important as upper values increase the risks for hypertension, diabetes, cardiovascular disease, gallstones, arthritis, and cancer. Skinfold measurement, also known as pinch test, is measured by a caliper at several precise points on the body and allows determining subcutaneous fat layer thickness and body fat percentage.

Table 3.6 Age-specific skinfold measures among the Garo women of Kamrup district, Assam

Skin fold Measures (mm)	*20 – 25 years N = 44 Mean ± SD*	*25 – 30 years N=39 Mean ± SD*	*30 – 35 years N=41 Mean ± SD*	*35 – 40 years N=40 Mean ± SD*	*40 – 45 years N=40 Mean ± SD*	*45 – 50 years N= 34 Mean ± SD*	*50 – 55 years N= 25 Mean ± SD*	*55 – 60 years N= 20 Mean ± SD*
Skin fold at Bicep	10.6 ± 0.63	11.3 ± 0.59	12.1 ± 0.59	12.7 ± 0.46	12.0 ± 0.51	12.0 ± 0.36	11.8 ± 3.52	11.8 ± 0.53
Skin fold at Tricep	12.9 ± 0.54	15.1 ± 0.51	15.6 ± 0.74	15.1 ± 0.47	15.7 ± 0.52	16.6 ± 0.49	14.1 ± 0.59	17.15 ± 0.65
Sub-Scapular Skin fold	15.4 ± 0.61	19.1 ± 0.81	20.3 ± 0.96	20.0 ± 0.69	20.4 ± 0.72	21.2 ± 0.74	19.2 ± 0.83	23.2 ± 0.90
Supra-Illiac Skin fold	18.6 ± 1.29	20.4 ± 1.34	20.4 ± 1.04	19.7 ± 0.72	17.6 ± 0.65	19.9 ± 0.67	17.6 ± 0.95	20.3 ± 1.00
Medial Calf Skin fold	17.4 ± 0.64	20.7 ± 0.82	21.1 ± 1.02	21.4 ± 0.92	19.9 ± 0.64	19.8 ± 0.46	19.6 ± 0.73	19.4 ± 0.78

Table 3.6 shows descriptive statistics pertaining to age-specific skin fold measurement at bicep, triceps, subscapular, supra-iliac and medial calf.

Table 3.7: Age-specific skin fold-based body composition indices among the Garo women of Kamrup district, Assam

Skin fold based body composition Indices	*20 – 25 years N = 44*	*25 – 30 years N=39*	*30 – 35 years N=41*	*35 – 40 years N=40*	*40 – 45 years N=40*	*45 – 50 years N= 34*	*50 – 55 years N= 25*	*55 – 60 years N= 20*
	Mean ± SD	***Mean ± SD***	***Mean ± SD***	***Mean ± SD***	***Mean ± SD***	***Mean ± SD***	***Mean ± SD***	***Mean ± SD***
Centripetal Fat Ratio	0.55± 0.09	0.55±0.05	0.55±0.77	0.56±0.07	0.57± 0.06	0.55± 0.04	0.57± 0.06	0.57± 0.09
Sub scapular- Triceps Skin fold Ratio (SSTSFR)	1.26 ± 1.23	1.27 ± 0.31	1.29 ± 0.34	1.34 ± 0.33	1.34 ± 0.30	1.37 ± 0.21	1.40 ± 0.38	1.42 ± 0.54
Sub scapular-Suprailiac-Tricep-Bicep Skin fold Ratio (SSI-TBSFR)	1.51 ± 0.78	1.51 ± 0.98	1.44 ± 0.39	1.45 ± 0.33	1.50 ± 0.08	1.53 ± 0.23	1.52 ± 0.31	1.52 ± 0.38
Sum of Trunkal Skin fold (STSF)	33.1 ± 1.55	37.7 ± 1.84	39.7 ± 1.95	39.8 ± 1.28	40.2 ± 1.27	41.2 ± 1.36	40.1 ± 1.71	41.4 ± 2.04
Sum of Peripheral Skin fold (SPSF)	40.9 ± 1.49	47.1 ± 1.67	47.4 ± 2.06	49.4 ± 1.51	49.5 ± 1.57	49.5 ± 1.15	49.7 ± 1.51	45.9 ± 1.99
Sum of Trunkal And Peripheral Skin fold Ratio (STPSFR)	0.81 ± 0.48	0.82 ± 0.43	0.83 ± 0.21	0.84 ± 0.17	0.88 ± 0.10	0.90 ± 0.12	0.92 ± 0.18	0.90 ± 0.20

As the age increases the skinfold thickness increases at all the sites except for bicep and medial calf skinfold. At biceps, the maximum mean value of skinfold measure is found in the age cohort of 35-40 years (12.7 mm). Same is for medial calf skinfold when between 35-40 years the maximum mean value (21.4 mm) is found. The higher fat percentage in women is due to sex-specific fat, such as breast, uterus and other sex-related fat deposits (Hoeger and Hoeger, 1999). He Ye in 2012, found that the subcutaneous fat is accumulated gently from the 20 age group in Wuweiin Gansu Province in China. And then, in the 40 age group, the thickness reaches a peak and it begins to thin after 40 year old group. The reason is that the age of the female skinfold thickness trends was related to the female hormone secretion. The percentage of essential body fat for females is greater than that for males, due to the demands of childbearing and other hormonal functions. After menopause (about 45-50 years old), the abilities of the female hormone secretion are drastically reduced, the abilities of the accumulation of fat get diminished, and the fat layer becomes thin (Suo et al., 2005). In the present study, such a possibility may not be ruled out but it needs a thorough study to say so.

The assessment of skinfold thickness is an objective measure of adiposity. Some adiposity-related indices are shown by age and sex in Table 3.7. Skinfold thicknesses measure subcutaneous fat at one or more sites to characterize total adiposity and they are restricted to the selected sites and do not measure internal fat (Moreno et al., 2007). The body fat percent of a human or other living being is the total mass of fat divided by total body mass; body fat includes essential body fat and storage body fat. Essential body fat at the same time is necessary to maintain life and reproductive functions. Centralization of body fat starts occurring after puberty. The centripetal fat ratio has been found to be more or less constant.

In the present study, it could be observed that the sum of truncal and peripheral skinfold ratio increases with the increase of age till 55 years of age but afterward it decreases. A similar trend of higher truncal and abdominal fat deposition with age, mainly in males were also found by Tresaco et al. (2009). There are several factors that influence body composition. Previous studies (Reid, 2002; Zhang, 2013) indicated that aging and estrogen levels of menopause are two important ones in postmenopausal women. Serum levels of sclerostin (Modder, 2011) may also play a key role. It was found by Jiang (2015) that age-related changes in body composition are complicated and needed to be confirmed by longitudinal studies. The present findings support these results that age-related changes do not show any particular trend which is evident from

Table 3.8: Age-specific upper arm compositions among the Garo women of Kamrup district, Assam

Upper arm variables	*20– 25 years N = 44*	*25 – 30 years N=39*	*30 – 35 years N=41*	*35 – 40 years N=40*	*40 – 45 years N=40*	*45 – 50 years N= 34*	*50 – 55 years N= 25*	*55 – 60 years N= 20*
	Mean ± SD	*Mean ± SD*	*Mean ± SD*	*Mean ± SD*	*Mean ± SD*	*Mean ± SD*	*Mean ± SD*	*Mean ± SD*
Total Upper Arm Area	46.68±10.58	52.88±10.64	57.12±11.93	53.56±12.44	51.09±14.39	54.81±6.24	52.82±11.72	54.62±10.65
Upper Arm Muscle Area	29.31 ± 6.28	34.51 ± 5.47	35.91 ± 8.18	35.07 ± 8.44	33.34 ± 7.86	34.60 ± 2.81	35.41 ± 5.29	34.20 ± 10.88
Upper Arm Fat Area	14.48 ± 6.86	17.89 ± 7.34	20.71 ± 11.28	21.98 ± 6.77	21.27 ± 7.92	19.71 ± 6.39	22.05 ± 10.04	25.93 ± 13.28
Upper Arm Muscle Area by Height	0.25 ± 0.35	0.22 ± 0.03	0.24 ± 0.05	0.17 ± 0.20	0.21 ± 0.04	0.23 ± 0.02	0.23 ± 0.03	0.22 ± 0.07
Mid Upper Arm Circumference	20.01 ± 1.96	20.90 ± 1.68	21.25 ± 2.39	20.99 ± 2.44	20.47 ± 2.34	20.97 ± 0.87	21.18 ± 1.57	20.24 ± 5.23

subscapular-suprailiac-tricep-bicep skinfold ratio. Neither the present nor previous studies could confirm a highly significant curvilinear relationship between age and fat mass. It has been noted that fat mass peaks in late middle age (Mott et al., 1999).

Age-specific upper arm composition among the Garo women has been presented in Table 3.8. The upper-arm composition is usually assessed using upper-arm muscle area (UMA), upper-arm fat area (UFA), arm-fat index (AFI), upper-arm fat area estimate (UFE) and upper-arm muscle area estimate (UME) calculated from mid-upper-arm circumference (MUAC) and triceps (TSF) skinfold thickness are used to determine the body-composition (Gibson, 2005). The skin folds simply derive to quantify the amount of muscularity and adiposity. In the present study no particular trend could be noticed in the total upper arm area after 35 years of age but till 35 years it increases.

Mid-arm muscle circumference (MAMC), or specifically mid-upper arm muscle circumference, can be used to evaluate fat-free mass or lean components of the body in nutritional assessment (Gassull et al., 1984). Lower values are associated with adverse outcome including increased risk of mortality in critical illness (Sungurte-kin et al., 2008), haemodialysis (Huang et al., 2010), HIV and tuberculosis infection (Villamor et al., 2006) and in people aged $\geq$ 80 years (Landi et al., 2010). Both MAMA and corrected MAMA is associated with clinical risk, including a prediction of the length of hospital stay in surgical patients (Almeida et al., 2013) and an increased risk of mortality in the elderly and those with chronic obstructive pulmonary disease (Miller et al., 2002); that MAMC reduces with age in men but not in women and that more recent data are greater, notwithstanding the difference in age banding, than the earlier values suggesting population change. No trend could be noticed in upper arm muscle area but fat area after 45 years it's fluctuating and before 45 years it increases.

Finally, although the present study shows some significant changes and trend with an increment of age in anthropometric and body composition traits, some variables fail to demonstrate a specific trend with an increment of age. From the above study, it can be concluded that stature decreases with age due to the shrinkage of inter-vertebral space and increased the curvature of the spine. With advancing age the muscles degenerate leading to the accumulation of fat. The prevalence of obesity increases with age showing 11.31 percent of latent obesity. A study showing age-specific changes (considering each age) would be more helpful, which could specifically be able to demonstrate the exact nature of change at the exact age.

Acknowledgements

Thanks are extended to all the villagers for their unswerving help in the data collection. We are indebted to them all.

References

Almeida, AI, Correira, M, Camilo, M and Ravasco, P 2013, 'Length of stay in surgical patients: nutritional predictive parameters revisited'. *British Journal of Nutrition,* Vol. 109, pp. 322–328.

Bell, AC, Ge, K and Popkin, BM 2002, 'The road to obesity or the path to prevention: motorized transportation and obesity in China'. *Obesity Research*, Vol. 10, pp. 277–283.

Bolzan, A, Guimarey, L and Frisancho, AR 1999, 'Study of growth in rural school children from Buenos Aires, Argentina using upper arm muscle area by height and other anthropometric dimensions of body composition'. *Annals of Human Biology,* Vol. 26, pp. 185–193.

Bujko, J, Jacek, Kasprzak, Houlshof, P and Schreurs, V 2006, 'Comparison of different methods of body fat measurement in non-obese young adults', *Polish Journal of Food and Nutrition Sciences,* Vol 15/56, SI 2, pp. 139-144.

Center for Disease Control and Prevention 2007, National Health and Nutrition Examination.

Das, BM and Roy, SK 2010, 'Age changes in the anthropometric and body composition characteristics of the Bishnupriya Manipuris of Cachar district, Assam'. *Advances in Bioscience and Biotechnology*, Vol. 1, no. 2, Article ID:2090.

Deepa, R, Sandeep, S and Mohan, V 2006, Abdominal obesity, visceral fat and type 2 diabetes- Asian Indian phenotype. Mohan V, Rao GHR, (eds.) *Type 2 diabetes in South Asians: epidemiology, risk factors, and prevention*. New Delhi, India: Jaypee Brothers Medical Publishers (P) Ltd; 2006. pp. 138-152.

Durnin, JV and Womersley, J 1974, 'Body fat assessed from total body density and skinfold thickness: measurements on 481men and women from 16 to 72 years'. *British Journal of Nutrition,* Vol. 32, pp. 77-97.

Engelhardt, M and Keller, K 2014, 'Strength and muscle mass loss with the aging process'. *Age and strength loss, Muscles Ligaments Tendons journal*, Vol. 3, no.4, pp. 346-350.

Frisanho, AR 1981, 'New norms of upper limb fat and muscle areas for assessment of nutritional status'. *American Journal of Clinical Nutrition*, Vol. 34, pp. 2540-2545.

Frisancho, AR 1989, *Anthropometric standard for the assessment of growth and nutrition status.* Ann Arbor, MI: University of Michigan Press.

Frisancho, AR and Tracer, DP 1987, 'Standards of arm muscle by stature for the assessment of nutritional status of children'. *American Journal of Physical Anthropology*, Vol. 73, pp. 459-465.

Gassull, MA, Cabré, E, Vilar, L, Alastrue, A and Montserrat, A 1984, *Protein-energy malnutrition: an integral approach and a simple new classification*. Human

Nutrition Clinical Nutrition, 38: Gibson RS 2005, Principle of nutritional assessment. Oxford: Oxford University Press. pp. 419-431.

Goran, MI 1998, 'Measurement issues related to studies of childhood obesity: activity, and food intake'. *Paediatrics*, Vol. 101, no.505.

Goran, MI, Gower, BA, Treuth, M and Nagy, TR 1998, 'Prediction of intra-abdominal and subcutaneous abdominal adipose tissue in healthy pre-pubertal children'. *International Journal of Obesity*, Vol. 22, pp. 549-558.

Haffner, SM, Diehl, AK, Stern, MP and Hazuda, HP 1989, 'Central adiposity and gallbladder disease in Mexican Americans'. *American Journal of Epidemiology*; Vol. 129, pp. 587-595.

He Ye, Hai Xiangjun, Bai Jingya, Wang Jing, He Jinquan, Wang Yutang and Xi Huanjiu 2012, 'A statistical analysis of skinfold thickness and measurement of body composition for Chinese adults'. *Scientific Research and Essays,* Vol. 7, no.31, pp. 2827-2834.

Heyward, Vivian H and Lisa M Stolarczyk 1996, 'Applied body composition assessment'. *Human Kinetics*.

Hoeger, W, and Hoeger, S 1999, *Principle and laboratory for physical fitness and wellness*. Englewood, Colorado: Morton Publishing Company.

Huang, Z, Willett, WC, Colditz, GA, Hunter, DJ, Manson, JE, Rosner, B, Speizer, FE and Hankinson, SE 1999, 'Waist circumference, waist-hip ratio, and risk of breast cancer in the Nurses' health study'. *American Journal of Epidemiology*, vol .150, no.12, pp. 1316-1324.

Huang, CX, Tighiouart, H, Beddhu, S, Cheung, AK, Dwyer, JT, Eknoyan, G, Beck, GJ, Levey, AS and Sarnak, MJ 2010, 'Both low muscle mass and low fat are associated with higher all-cause mortality in hemodialysis patients'. *Kidney International*, Vol. 77, pp. 624-629.

Hulsegge, G, Herber-Gast, GC, Spijkerman, AM, Susan, H, Picavet, J, Van der Schouw, Bakker, SJ, Gansevoort, RT, Dollé, ME, Smit, HA, Monique and Verschuren, WM 2016, 'Obesity and age-related changes in markers of oxidative stress and inflammation across four generations'. *Obesity (Silver Spring),* Vol. 24, no.6, pp. 1389-1396.

Huxley, R, Mendis, S, Zheleznyakov, E, Reddy, S and Chan, J 2010, 'Body mass index, waist circumference and waist-hip ratio as predictors of cardiovascular risk—a review of the literature'. *European Journal of Clinical Nutrition*, Vol. 64, pp. 16-22.

Joshi, SR 2003, 'Metabolic syndrome- Emerging clusters of the Indian phenotype (editorial)'. *The Journal of the Association of Physicians of India,* Vol. 51, pp. 445-446.

Jiang, Y, Zhang, Y, Jin, M, Gu, Z, Pei, Y and Meng, P 2015, 'Age-related changes in body composition and association between body composition with bone mass density by body mass index in Chinese Han Men over 50-year-old'. *Plos One,* Vol. 10, no.6.

Jylhava, J, Pedersen, NL and Hagg, S 2017, 'Biological age predictors'. *EBioMedicine,* Vol. 21, pp. 29-36.

Keska, A, Tkaczyk, J, Czajkowska, A, Wiœniewski, A, Norkowski, H, Smolarczyk, M and Kapuscinski, P 2012, 'Fat content in young adults determined by skinfolds and body composition analyzer'. *Pediatric Endocrinology, Diabetes, and Metabolism,* Vol. 18, no.1, pp. 33-36.

Kissebah, AH, Vydelingum, N, Murray, R, Evans, DJ, Hartz, AJ, Kalkhoff, RK and Adams, PW 1982, 'Relation of body fat distribution to metabolic complications of obesity'. *The Journal of Clinical Endocrinology and Metabolism,* Vol. 54, no.2, pp. 254-260.

Kravitz, Len, and Vivian H Heyward, 2018 *Body composition assessment–UNM edu* https://www.unm.edu/~lkravitz/Article%20folder/bodycomp.html. Assessed on 3rd May 2018.

Kushner, RF, Kunigk, A, Alspaugh, M, Andronis, PT, Leitch, CA and Schoeller, DA 1990, 'Validation of bioelectrical impedance analysis as a measurement of change in body composition in obesity'. *Asia Pacific Journal of Clinical Nutrition*, Vol. 52, no.2, pp. 219-223.

Kyle, UG, Bosaeus, I, De Lorenzo, AD, Deurenberg, P, Elia, M, Manuel Gómez, J, Lilienthal Heitmann,B, Kent-Smith, L, Melchior, JC, Pirlich, M, Scharfetter, H, MWJ Schols, A and Pichard, C 2004, 'Bioelectrical impedance analysis-part II: utilization in clinical practice'. *Clinical Nutrition*, Vol. 23, no.6, pp. 1430-1453.

Landi, F, Russo, A, Liperoti,R, Pahor, M, Tosato, M, Capoluongo, E, Bernabei, R and Onder G 2010, 'Mid-arm muscle circumference, physical performance,and mortality: results from the aging and longevity study in the Sirente geographic area (ilSIRENTE study)'. *Clinical Nutrition*, Vol. 29, pp. 441-447.

Lexell, J, Taylor, CC, and Sjostrom, M 1988, 'What is the cause of the ageing atrophy? Total number, size, and proportion of different fiber types studied in whole vastus lateralis muscle from 15 to 83-year-old men.' *Journal of Neurology Science,* Vol. 84, pp. 275-294.

Liu, J, Fox, CS, Hickson, DA, May, WD, Hairston, KG, Carr, JJ and Taylor, HA 2010, 'Impact of abdominal visceral and subcutaneous adipose tissue on cardiometabolic risk factors: the Jackson heart study'. *The Journal of Clinical Endocrinology and Metabolism,* Vol. 95, no.12, pp. 5419-5426.

Lowsky, DJ, Olshansky, SJ, Bhattacharya, J and Goldman, DP 2017, 'Heterogeneity in healthy aging'. *The Journals of Gerontology*. Series A, Biological Sciences, and Medical Sciences, Vol. 69, no.6, pp. 640–649.

Miller, M D, Crotty, M, Giles, LC, Bannerman, E, Whitehead, C, Cobiac L, Daniels, LA and Andrews, G 2002, 'Corrected arm muscle area: an independent predictor of long-term mortality in community-dwelling older adults?' *Journal of the American Geriatrics Society,* Vol. 50, pp. 1272-1277.

Mödder, UI, Hoey, KA, Amin, S, McCready, LK, Achenbach, SJ, Riggs, BL, Melton, LI 3rd, Khosla, S. 2011, 'Relation of age, gender, and bone mass to circulating sclerostin levels in women and men'. *Journal of Bone Mineral Research*, Vol. 26, no.2, pp. 373-379.

Moreno, L.A., Mesana, M.I., González-Gross, M., Gil, C.M., Ortega, F.B., Fleta, J., Wärnberg, J., León, J, Marcos, A and Bueno, M 2007, 'Body fat distribution reference standards in Spanish adolescents: The AVENA Study'. *International Journal of Obesity (Lond.),* Vol. 31, no.12, pp. 1798-1805.

Mott, JW, Wang, J, Thornton, JC, Allison, DB, Heymsfield, SB and Pierson, RN Jr 1999, 'Relation between body fat and age in 4 ethnic groups'. *American Journal of Clinical Nutrition*, Vol. 69, no. 5, pp. 1007-1013.

Misra, A, Singhal, N and Khurana, L 2010, 'Obesity, the metabolic syndrome, and type 2 diabetes in developing countries: the role of dietary fats and oils'. *Journal of the American College of Nutrition*, Vol. 29(Suppl 3), pp. 289S-301S.

National Institutes of Health 1995, *Bioelectrical impedance analysis in body composition measurement*. National Institute of Health Technology Assessment Conference Statement.

Ode, J J, Pivarnik, J M, Reeves, M J and Knous, J L 2007, 'Body mass index as a predictor of percent fat in college athletes and nonathletes'. *Medicine and Science in Sports and Exercise,* Vol. 39, no. 3, pp. 403-409.

Osei, K, Cottrell, DA and Orabella, MM 1991, 'Insulin sensitivity, glucose effectiveness, and body fat distribution pattern in nondiabetic offspring of patients with NIDDM', *Diabetes Care,* Vol. 14, pp. 890-896.

Pradeepa, R, Anjana, RM, Joshi, SR, Bhansali, A, Deepa,M, Joshi, PP, Dhandania,VK, Madhu, V, Rao, PV, Geeta, L, Subashini, R, Unnikrishnan, R,Kaur, T, Mohan, V, Das AK and ICMR-INDIAB Collaborative Study Group 2015, 'Prevalence of generalized and abdominal obesity in urban and rural India- the ICMR - INDIAB Study (Phase-I) [ICMR - INDIAB-3]'. *Indian Journal of Medical Research*, Vol. 142, no. 2, pp. 139-150.

Reid, IR 2002, 'Relationships among body mass, its components, and bone'. *Bone* (NY), Vol. 31, pp. 547-555.

Ricciardi, R and Talbot, LA 2007, 'Use of bioelectrical impedance analysis in the evaluation, treatment, and prevention of overweight and obesity'. *Journal of the American Academy of Nurse Practitioners,* Vol. 19, pp. 235-241.

Roy, SK and Pal, B 2003, 'Anthropometric and physiological traits: Age changes among the Oraon agricultural labourer of the Jalpaiguri district, Northern West Bengal, India'. *Anthropology Anzeiger*, Vol. 61, no. 4, pp. 445-460.

Sehested, TS, Hansen, TW, Olsen, MH, Abildstrøm, SZ, Ramussen, S, Ibsen, H, Torp-Pedersen, C, Madsbad, S and Jeppesen, J 2010, 'Measures of overweight and obesity and risk of cardiovascular disease: a population-based study'. *European Journal of Cardiovascular Prevention and Rehabilitation*, Vol. 17, no. 4, pp. 486-490.

Sungurtekin, H, Sungurtekin, U, Oner, O and Okke, D 2008, 'Nutritional assessment in critically ill patients'. *Nutrition in Clinical Practise*, Vol. 23, pp. 635-641.

Suo, L, Lu S, Zhang 2005, 'Skinfold thickness and its variations in RUSS nationality adults.' *Journal of Harbin Normal University (Natural science),* Vol. 21, no.3, pp. 103-108.

Thibault, R, Genton, L and Pichard, C 2012, 'Body composition: why, when and for who?' *Clinical Nutrition*, Vol. 31, pp. 435-447.

Thompson, CJ, Ryu, JE, Craven, TE, Kahl, FR and Crouse, JR 3rd 1991, 'Central adipose distribution is related to coronary atherosclerosis'. *Arteriosclerosis and Thrombosis,* Vol. 11, no. 2, pp. 327-333.

Tresaco, B, Moreno, LA, Ruiz, JR, Ortega, FB, Bueno, G, González-Gross, M, Wernberg, J, Gutiérrez, A, García-Fuentes, M, Marcos, A, Castillo, MJ and Bueno, M 2009, 'Truncal and abdominal fat as determinants of high triglycerides and low HDLcholesterol in adolescents'. *Obesity,* Vol. 17, no.5, pp. 1086-1091.

Villamor, E, Saathoff, E, Mugusi, F, Bosch, RJ, Urassa, W and Fawzi, WW 2006, 'Wasting and body composition of adults with pulmonary tuberculosis in relation to HIV-1 coinfection, socio-economic status, and severity of tuberculosis'. *European Journal of Clinical Nutrition,* Vol. 60, pp. 163-171.

Visscher, TL, Seidell, JC, Molarius, A, Van der Kuip, D, Hofman, A and Witteman, JC 2001, 'A comparison of body mass index, waist-hip ratio and waist circumference as predictors of all-cause mortality among the elderly: the rotterdam study'. *International Journal of Obesity and Related Metabolic Disorders*, Vol. 25, no.11, pp. 1730-1735.

Wagner, DR, and Heyward, VH 1999, *Techniques of body composition assessment: a review of laboratory and field methods*. Research Quarterly for Exercise and Sports Survey. Anthropometric Procedures Manual. Hyattsville: CDC. Vol. 70, pp. 135-149.

Wannamethee, SG, Shaper, AG and Lennon, L 2005, 'Reasons for intentional weight loss, unintentional weight loss, and mortality in older men'. *Archives of Internal Medicine*, Vol. 165, pp. 1035-1040.

World Health Organisation 1995, *Physical status: The use and interpretation of anthropometry. Report of a WHO expert committee technical report series*, No. 854.

World Health Organization 2003, *Diet, nutrition and the prevention of chronic diseases.Report of a joint WHO/FAO expert consultation. World health organ tech rep series,* no. 916: i–viii.

World Health Organization 2008, *Waist circumference and waist-hip ratio: Report of a WHO expert consultation,* Geneva, 8–11 December 2008.

World Health Organization 2009, *Global health risks: mortality and burden of disease attributable to selected major risks*. Geneva: Switzerland.

Zhang, H, Chai, X, Li, S, Zhang, Z, Yuan, L, Xie, H, Zhou, H, Wu, X, Sheng, Z and Liao, E 2013, 'Age-related changes in body composition and their relationship with bone mineral density decreasing rates in central south Chinese postmenopausal women'. *Endocrine*, Vol. 43, no.3, pp. 643-650.

Chapter - 4

Nutritional Status and Prevalence of Anemia in Two Communities of Cachar District in Assam, India

—A. F. Gulenur Islam Barbhuiya and Rekha Das

Introduction

The importance of malnutrition as well as anemia as a primary health problem throughout the globe is extensively recognized. Although there have been impressive improvements in some health indicators in the last two decades, including a reduction in infant mortality rate and a drop in the fertility rate, improvements in nutritional status, have been less impressive (IIPS, 2007). More than one-third of Indian adult population has a body mass index (BMI) of less than 18.5 kg/m^2, which reflects the chronic energy and micronutrient deficit (Bentley and Griffiths, 2003). National Family Health Survey (NFHS-3) revealed some of the important facts about the status of malnutrition and anemia among the adult population of India. Adults in India suffer from a dual burden of malnutrition (abnormal thinness and overweight or obesity). 36 percent of women and 34 percent of men are undernourished, with a BMI less than 18.5, indicating a high prevalence of nutritional deficiency. Only 52 percent of women and 57 percent of men are at a normal weight for their height. Besides that more than half of the women (55.0%) and almost one-quarter of men (24.0%) are anemic (IIPS, 2007).

Anemia, another major public health problem, contributes to the vicious circle of undernutrition, to the detriment of already underprivileged groups in the developing countries (Bhattacherjee et al., 2010). In the case of both male and female, the prevalence of anemia was the highest among those with severe undernutrition (WHO, 2000). A hematological parameter like haemoglobin (Hb) is generally known to be affected by socio-economic conditions, especially via nutrition, as the nutritional status largely depends upon the socio-economic condition (Bentley and Griffiths, 2003; Bharati et al., 2004; Bhargava et al., 2001). Despite the knowledge that discrimination in nutrition and health care may lead to severe growth faltering and increased mortality, people tend to discriminate (World Bank, 1991). Findings of the National Family Health Survey (NFHS-2) data also indicated that subject with a low BMI had a somewhat higher prevalence of anemia than others (IIPS, 2002). Therefore, an attempt has been made to focus on the association between nutritional status (body mass index) and anemia (haemoglobin concentration level).

Objective

The objective of the present study is to see the nutritional status in relation to anemia among the adult Bengali Muslims and adult Meiteis.

The Relevance of the Study

Undernutrition and anaemia remain a major public health problem in nearly all developing countries, including India. Very few ethnic-based studies have been undertaken on the adult nutritional status and anemia from Northeast India (Basu, 2006; Biswas and Baruah, 2014; Haloi and Limbu, 2013; Khongsdier, 2001; Khongsdier and Basu, 2004; Medhi et al., 2006; Singh and Mangang, 2012). Apart from that, there is a lack of published data dealing with community-based studies on adult nutritional status and anemia from the southern part of Assam, India. The study tried to get a glimpse of the nutritional status and anemic situation among the adult Bengali Muslim and Meitei population. Moreover, the study had covered the working age group population who are generally considered as the backbone for the development of any society or community. Working-age group population is generally viewed as the population who are involved in different income generating as well as productive activities. Work capacity may determine work output and hence income; but income may also affect dietary intakes and ultimately work capacity (Ghosh and Bharati, 2003). Under-nutrition along with anemia among the adults may account for a significant loss of productivity and ultimately family welfare (Leslie, 1997). Besides, the study had also covered the women of reproductive and childbearing age group. Anaemia in childbearing female increases the risk of maternal morbidity and mortality, perinatal and neonatal mortality, inadequate iron stores for the newborn, premature delivery and low birth weight (Pal et al., 2014). If the women of that age group of a community are suffering from both malnutrition and anemia, children born out of these women are also bound to be affected.

Materials and Methods

The present study was done in four randomly selected villages (neighbouring) of Cachar district of Assam, primarily inhabited by Bengali Muslim and Meitei population. The study was conducted among the Bengali Muslims of Bhaurikandi Part-II and the Meiteis of Sundari Part-IV village of Sonai Block of Cachar District, Assam. On the other hand, the study was done among the Bengali Muslims of Ganganagar Part-I village and the Meiteis of Saint Katherine village of Palonghat block of Cachar District, Assam. General demographic and socio-economic data which includes relevant information on sex, age, education, occupation, house type, the source of drinking water, treatment of drinking water, sanitation, electrification, energy sources for cooking, family type, etc. were collected by following observation and interview method.

Anthropometric measurements like weight and stature as well as haemoglobin concentration level in the blood (g/dl) was done by Haemoglobin Colour Scale (HCS) method from 549 adults of 20 years to 69 years of age. The data had been collected from 268 Bengali Muslims (124 males and 144 females) and 281 Meiteis (131 males and 150 females). It was a cross-sectional study and the data were collected during the period of 2010-2011. The sample was free from any selection bias. All the available willing individuals who fall within the age group of 20 years to 69 years were included in the study. Pregnant women were excluded from the study.

Weight and height had been measured by using standard techniques (Weiner and Lourie, 1981). Height and weight had been recorded to the nearest 0.1 cm and 0.5 kg, using standard anthropometer and weighing scale respectively. Nutritional status of the adults was assessed from BMI by following WHO classification (WHO, 1995). Anemia status was assessed from haemoglobin levels (g/dl) by matching the colour of a blood drop on a test strip with one of the shades of red of colour scale which represent a range of haemoglobin values from 4 to 14 g/dl. The measured haemoglobin concentrations were used to classify individuals as severely, moderately, mildly anemic and non-anemic or normal individuals. Different grades of anemia were assessed from haemoglobin levels by following the WHO classification (WHO, 2001).

All statistical analysis was carried out by SPSS 16.0 version. Frequency, descriptive statistics, independent t-test, chi-square test, and Pearson's correlation coefficient were performed wherever required. A *p* value of <0.05 had been considered as significant in the required statistical tests.

The Land and the People

Cachar District at a Glance

Geographically Cachar district is one of the largest districts, located in the southern part of Assam. Although the district is considered under plain area there are a number of hills spread across the district. The climate is tropical wet in nature where summers are hot and wet and winters are cool (http://cachar.gov.in). It has two subdivisions, five Revenue Circles, fifteen Development Blocks, two towns, 163 Gaon Panchayats and 895 Revenue Villages (http://cachar.nic.in).

According to the 2011 census Cachar district has a population of 17,36,319 which gives it a ranking of 278th in India out of a total of 640. The district has a population density of 459 inhabitants per square kilometer. Cachar has a sex ratio of 958 females for every 1000 males and a literacy rate of 80.36 percent. The religious break-up of the population is Hindus 886,761, Muslims 522,051 and Christians 31,306 (District Census, 2011).

The Villages

The present study was done among the Bengali Muslims of Bhaurikandi Part-II and the Meiteis of Sundari Part-IV village of Sonai Block of Cachar District, Assam. On the other hand, the study was done among the Bengali Muslims of Ganganagar Part-I village and the Meiteis of Saint Katherine village of Palonghat block of Cachar District, Assam. The villages like Bhaurikandi Part-II, Sundari Part-IV, Saint Katherine and Gonganagar Part-I is situated at about 25, 32, 34 and 37 kilometers away respectively from Silchar town (district headquarter) in the south-eastern direction. The climate is tropical with hot and wet summer and cool winter which is more or less similar to those of the plain areas of other parts of Assam. The rainfall of the area is average or medium like the whole district. All the villages are full of floras and faunas. Silchar to Aizwal National Highway (54 No.) passes through Kabuganj junction from where district roads connected the villages to the head quarters of the district (Silchar).

There are lower primary schools and upper primary schools in all the villages. Apart from that high schools and higher secondary schools are also available for the students of all the villages in neighbouring areas. For higher studies, the students of all the villages go to Janata College, Kabuganj or Madhab Chandra Das College, Sonai. Mosques with burial grounds near each mosque are present in the two Bengali Muslim villages. Whereas temples, as well as cremation grounds, are also present in the two Meitei inhabited villages. Apart from these some of the Meitei houses have their own places of worship. Hathikhal Bazar and Kochudaram Bazar are the main marketplaces for both Bhaurikandi Part-II and Dakshin Mohanpur Part-V village. Amraghat Bazar is the important marketplace for both Ganganagar Part-I and Saint Katherine village while Darmi Bazar is the main marketplace for both Saint Katherine and Sundari Part-IV village. There are permanent shops and business establishments dealing with clothes, grocery, hardware, stationery, medicine, electrical items, etc. in the above-mentioned markets.

The house types that are present in all the villages are predominantly Assam type i.e. brick or bamboo wall with *tin* or C.I. sheet. The houses are arranged in a linear pattern, i.e. by both the sides of the village lane and bye-lanes. The main source of drinking water is the water supplied by state public health and engineering department (PHED) but some villagers use to drink water from the pond and dug well also. Although the drainage system in all the villages is not so good house sewage naturally flow towards low land as almost all the houses are prepared in a comparatively high land. The pit latrine is predominantly observed in the houses of all the villages. Electricity facility is available in the majority of the houses of all the villages. Firewood is the primary source for cooking among the people of the villages.

People

There are different endogamous communities inhabiting the district such as Bengali Hindu, Bengali Muslim, Meitei, Bishnupriya, Dimasa Kachari, Hmar, Khasi, etc. The people inhabiting in Barak valley (which includes Cachar, Karimganj and Hailakandi district) are primarily known as *Sylheti Bengali* (*Sylheti*, a dialect) who mainly follow either Islam or Hinduism. The present study took into consideration two of the largest communities of Cachar district namely Bengali Muslim and Meitei, residing as neighbours in most of the areas of the district. The valley has a long history of Islam as half of the valley came under the rule of the Turk-Afghan dynasties of Bengal from the early 14th century and continued with the establishment of the Mughal Empire ("Cachar district,' 'n.d.; Saikia, 2004). Meitei people got its introduction to the valley from the early part of the nineteenth century when its princes made Cachar a springboard for the reinvasion of the territory after being displaced from Meitrabak, modern Manipur ("Cachar district,' 'n.d.). Linguistically the Bengali Muslims belong to the Indo-European ethnic group of Caucasoid racial stock while the Meiteis belong to the Tibeto-Burman ethnic group of Mongoloid racial stock (Basu et al., 2005).

The Bengali Muslims speak in Sylheti Bengali(a dialect) while the Meiteis speak in Manipuri language. The Bengali Muslims belong to Sunni sect of Islam while the Meiteis follow *Gorio Baishnab Dharma,* a sect of Hinduism. Both the communities follow the patriarchal system of family structure, where the property is inherited through male lines. Although both nuclear and joint families are present in both the communities but nuclear families are more common among the Muslims. Monogamy is the prevalent form of marriage among the Muslims and the Meiteis of the area. But polygamy is not totally wiped out among the Muslims. Although marriage by negotiation is the prevailing practice in both the communities but elopement marriage is also observed especially among the Meiteis. Agriculture is the mainstay of livelihood for both the Muslims and the Meiteis. In both the communities, although a number of peoples are engaged in different occupations almost every household has one type or two types of wet cultivation such as *Aush* (Ahu- April/May to July/August) and *Hail* (Sali- May/June to November/December). Rice is the staple food of both the communities. The Muslims are generally non-vegetarian while most of the Meiteis are vegetarian. The habit of chewing betel nut and betel leaf is very much common, especially among the Muslims.

Results

Different grades of anemia have been observed on the basis of the nutritional status of the Muslims and the Meiteis (Table 4.1a, Table 4.1b and Figure 4.1). It is observed that 35.4 percent and 20.8 percent of the underweight Muslim males are moderately and mildly anemic respectively. Whereas moderate (19.4%) and mild (16.7%) anemia are found to be the lowest among the Muslim males who are having normal nutritional status (BMI: 18.5-25.0 in kg/m^2). Half of the overweight males are mildly anemic (50.0%) and the rest half is having normal haemoglobin status. On the other hand, 6.9 percent of the underweight Muslim females are severely anemic. 36.2 percent and 27.6 percent of the underweight Muslim females are found to be moderately and mildly anemic respectively. Although 32.4 percent of the moderately anemic Muslim females are having healthy nutritional status a large number of them are fall under in BMI category of 18.5 to 20.0 in kg/m^2. All overweight Muslim females are found to have normal haemoglobin status.

Table 4.1a: Different grades of anemia on the basis of the nutritional status of the Muslims

Com-munity	*Sex*	*Nutritional Status*		*Severe*	*Moderate*	*Mild*	*Normal*	*Total*
Muslim	Male	Under Weight		-	17 (35.4)	10 (20.8)	21 (43.8)	48 (100.0)
		Normal	1	-	9 (30.0)	3 (10.0)	18 (60.0)	30 (100.0)
			2	-	5 (11.9)	9 (21.4)	28 (66.7)	42
								(100.0)
			Total		14 (19.4)	12 (16.7)	46 (63.9)	72 (100.0)
		Over Weight		-	-	2 (50.0)	2 (50.0)	4 (100.0)
	Female	Under Weight		4 (6.9)	21 (36.2)	16 (27.6)	17 (29.3)	58 (100.0)
		Normal	1	2 (5.4)	12 (32.4)	8 (21.6)	15 (40.5)	37 (100.0)
			2	-	8 (17.8)	10 (22.2)	27 (60.0)	45 (100.0)
			Total	2 (2.4)	20 (24.4)	18 (22.0)	42 (51.2)	82 (100.0)
		Over Weight		-	-	-	4(100.0)	4 (100.0)
	Total	Under Weight		4 (3.8)	38 (35.8)	26 (24.5)	38 (35.8)	106 (100.0)
		Normal	1	2 (3.0)	21 (31.3)	11 (16.4)	33 (49.3)	67 (100.0)
			2	-	13 (14.9)	19 (21.8)	55 (63.2)	87 (100.0)
			Total	2 (1.3)	34 (22.1)	30 (19.5)	88 (57.1)	154 (100.0)
		Over Weight		-	-	2 (25.0)	6 (75.0)	8 (100.0)

Normal: 1:- BMI:18.5-19.99, 2:-BMI:20.0-24.99; In bracket the figure shows percentage

Moderate anemia is found to be very high in underweight (30.0%) status compared to normal (BMI: 18.5-25.0 in kg/m^2) nutritional status (8.4%) among the Meitei males. But the frequency of mild anemia is almost the same among the underweight (30.0%) and normal (30.1%) Meitei males. 37.5 percent of the overweight Meitei males are found to have mild anemia. The frequency of moderate anemia is found to be very high among the underweight (36.2%) Meitei females compared to those who are having healthy nutritional status (21.5%). Mildly anemic Meitei females are also found to be more in underweight (36.2%) category compared to normal

Table 4.1b: Different grades of anemia on the basis of the nutritional status of the Meiteis

Community	*Sex*	*Nutritional Status*		*Severe*	*Moderate*	*Mild*	*Normal*	*Total*
Meitei	Male	Under Weight		-	12 (30.0)	12 (30.0)	16 (40.0)	40 (100.0)
		Normal	1	-	6 (20.0)	7 (23.3)	17 (56.7)	30 (100.0)
			2	-	1 (1.9)	18 (34.0)	34 (64.2)	53 (100.0)
			Total		7 (8.4)	25 (30.1)	51 (61.4)	83 (100.0)
		Over Weight		-	-	3 (37.5)	5 (62.5)	8 (100.0)
	Female	Under Weight		1 (2.1)	17 (36.2)	17 (36.2)	12 (25.5)	47 (100.0)
		Normal	1	2 (4.8)	9 (26.2)	17 (35.7)	14 (33.3)	42 (100.0)
			2	-	11 (17.6)	13 (31.4)	26 (51.0)	51 (100.0)
			Total	2 (2.2)	20 (21.5)	31 (33.3)	40 (43.0)	93 (100.0)
		Over Weight		-	-	2 (25.0)	6 (75.0)	8 (100.0)
		Obesity		-	-	2 (100.0)	-	2 (100.0)
	Total	Under Weight		1 (1.1)	29 (33.3)	29 (33.3)	28 (32.2)	87 (100.0)
		Normal	1	2 (2.8)	17 (23.6)	22 (30.6)	31 (43.1)	72 (100.0)
			2	-	10 (9.6)	34 (32.7)	60 (57.7)	104 (100.0)
			Total	2 (1.1)	27 (15.3)	56 (31.8)	91 (51.7)	176 (100.0)
		Over Weight		-	-	7 (38.9)	11 (61.1)	18 (100.0)

Normal: 1:- BMI:18.5-19.99,2:-BMI:20.0-24.99; In bracket the figure shows percentage

(33.3%) status. It is notable that two Meitei women who are having better nutritional status are found to be severely anemic but both of them have BMI value of 18.5-20.0 in kg/m^2. Only mild anemia is observed among the overweight (25.0%) and obese (100.0%) females of the Meitei population.

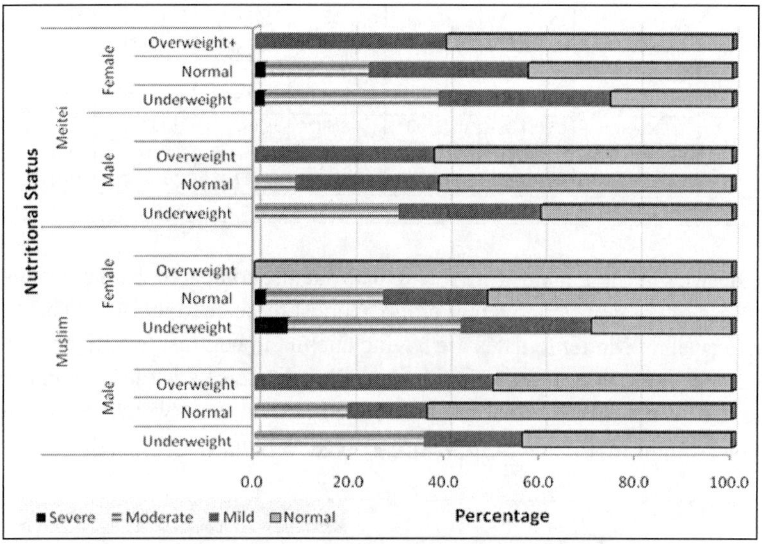

Figure 4.1: Different grades of anemia on the basis of the nutritional status of the Muslims and the Meiteis.

Prevalence of anemia has been observed in relation to underweight (BMI: <18.5 in kg/m^2) among the Muslims and the Meiteis by compiling the BMI value of ≥18.5 (in kg/m^2) into one category and termed as non-underweight (Table 4.2 and Figure 4.2). It is observed that 64.2 percent of the Muslims are anemic in the underweight category compared to 42.0 percent anemia in the non-underweight group. Sex wise distribution of anemia also shows that 56.2 percent of the underweight Muslim males are anemic in contrast to 36.8 percent in the non-underweight group. Anemia is also found to be more in underweight (70.0%) category compared to non-underweight (46.5%) group among the Muslim females. There are statistically significant differences between the underweight and non-underweight group in the prevalence of anemia among the Muslim males (*p*-0.042) as well as females (*p*-0.006).

On the other hand, the frequency of anemia is also found to be very high (67.8%) in the underweight group compared to non-underweight (47.4%) group among the Meiteis. More than half (60.0%) of the

Table 4.2: Prevalence of anemia among the underweight Muslims and Meiteis

Sex	Nutritional Status	Muslim			Meitei		
		Anemic	*Normal*	*Chi-Square*	*Anemic*	*Normal*	*Chi-Square*
Male	Underweight	27 (56.2)	21 (43.8)	χ^2-4.490 d.f.-1 *p*-0.042	24 (60.0)	16 (40.0)	χ^2-5.207 d.f.-1 *p*-0.035
	Non-underweight	28 (36.8)	48 (63.2)		35 (38.5)	56 (61.5)	
Female	Underweight	41 (70.7)	17 (29.3)	χ^2-8.228 d.f.-1 *p*-0.006	35 (74.5)	12 (25.5)	χ^2-4.979 d.f.-1 *p*-0.030
	Non-underweight	40 (46.5)	46 (53.5)		57 (55.3)	46 (44.7)	
Total	Underweight	68 (64.2)	38 (35.8)	χ^2-12.607 d.f.-1 *p*-0.000	59 (67.8)	28 (32.2)	χ^2-10.048 d.f.-1 *p*-0.002
	Non-underweight	68 (42.0)	94 (58.0)		92 (47.4)	103(52.6)	
In bracket the figure shows percentage							

underweight Meitei males are found to be anemic compared to 38.5 percent in case of the non-underweight group. Almost three quarter (74.5%) of the underweight Meitei females are having anemia in contrast to 55.3 percent anemia in the non-underweight group. Prevalence of anemia is found to have a statistically significant difference between underweight and non-underweight Meitei males (*p*-0.035) as well as females (*p*-0.030).

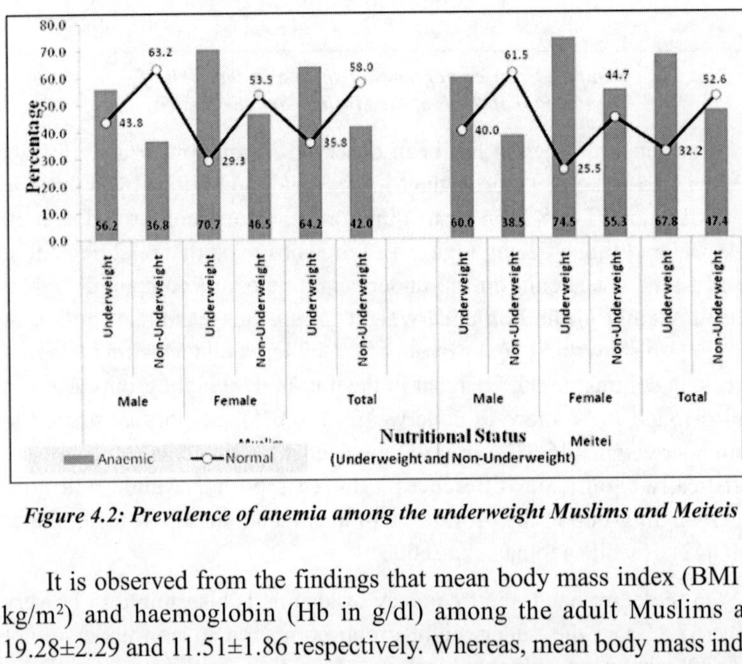

Figure 4.2: Prevalence of anemia among the underweight Muslims and Meiteis

It is observed from the findings that mean body mass index (BMI in kg/m^2) and haemoglobin (Hb in g/dl) among the adult Muslims are 19.28±2.29 and 11.51±1.86 respectively. Whereas, mean body mass index and haemoglobin among the adult Meiteis are 20.12±2.75 and 11.81±1.58

respectively. Independent t-test reveals that mean BMI is significantly higher (*p*: 0.000) among the Meiteis compared to the Muslims. However, it is noticed from the same test that means haemoglobin of the Meitei community is almost significantly higher (*p*: 0.050) in contrast to that of the Muslim community (Table 4.3). Mean body mass index and mean haemoglobin give more or less similar picture when it is observed on the basis of the sex of the two communities.

Table 4.3: Correlation between body mass index and haemoglobin level along with average values among the Muslims and the Meiteis

Variables			*Body Mass Index (BMI in Kg/m²)*			*Haemoglobin (Hb in g/dl)*			*Pearson correlation coefficient between BMI and Hb*
Community	*Sex*	*No.*	*Mean*	*SD*	*SE*	*Mean*	*SD*	*SE*	
Muslim	Male	124	19.39	2.16	0.19	12.07	1.78	0.16	0.283**
	Female	144	19.19	2.39	0.20	11.04	1.80	0.16	0.393**
	Total	268	19.28	2.29	0.14	11.51	1.86	0.16	0.343**
Meitei	Male	131	20.21	2.53	0.22	12.30	1.48	0.13	0.340**
	Female	150	20.04	2.94	0.24	11.37	1.54	0.13	0.198*
	Total	281	20.12	2.75	0.16	11.81	1.58	0.09	0.255**
Indepen -dent t-test (Between)	Male and Female (Muslim)		t: 6.809, df: 266, *p*: 0.485			t: 4.702, df: 266, *p*: 0.000			Correlation (2-tailed) is significant at the: *0.05 level and **0.01 level
	Male and Female (Meitei)		t: 0.522, df: 279, *p*: 0.602			t: 5.158, df: 279, *p*: 0.000			
	Muslim and Meitei (Male)		t: 2.796, df: 253, *p*: 0.006			t: 1.138, df: 253, *p*: 0.256			
	Muslim and Meitei (Female)		t: 2.719, df: 292, *p*: 0.007			t: 1.702, df: 292, *p*: 0.090			
	Muslim and Meitei (Total)		t: 3.883, df: 547, *p*: 0.000			t: 1.968, df: 547, *p*: 0.050			

Pearson's correlation coefficient discloses that (Table 4.3) body mass index (BMI in kg/m²) of the adult Muslims are found to have a significant positive correlation (r_{xy}=0.343, p<0.01) with their hemoglobin concentration level (Hb in g/dl). Similar results are also noticed in both the sexes of the Muslim community (male: r_{xy}=0.283, p<0.01; female: r_{xy}=0.383, p<0.01). On the other hand like the Muslims, body mass index of the adult Meiteis are also found to have a significant positive correlation with their hemoglobin (r_{xy}=0.255, p<0.01). The same trend is also visible among the males (r_{xy}=0.340, p<0.01) and females (r_{xy}=0.198, p<0.05) of the Meitei community.

Discussion

The study reveals that the severity of anemia increases with the decrease in the level of BMI in both communities. Sex wise distribution of different grades of anemia also gives a similar picture. The findings of the present study are also in conformity with some earlier researches carried out among rural adult males and non-pregnant females of India focusing on the association between anemia and nutritional status based on BMI. Ramachandra and Kasthuri (2008) demonstrated the association between the higher prevalence of anemia and a low BMI in their study among the elderly South Indian rural population. Low BMI and anemia were found to be present at considerable levels in a study by Bhattacharjee et al. (2010) among the adult males and females living in rural Coochbehar, North Bengal. They also demonstrated that with the increase in body mass index, haemoglobin level also increases significantly. Presence of severe and moderate anemia is very high in underweight individuals of both the communities in contrast to those individuals who are having healthier nutritional status, whereas the absence of both severe and moderate anemia is noticed among the overweight or obese individuals of both the communities. Individuals having mild anemia show a fluctuating trend from underweight to overweight and obesity through normal nutritional status. In the present study, special emphasis has been given on the presence of anemia among the underweight individuals (or association between anemia and underweight) as large sections of the population of both the communities are anemic as well as underweight. Presence of anemia is found to be significantly high among underweight individuals of both the communities compared to the non-underweight individuals. The results also indicate that underweight women of both communities are more vulnerable to anemia compared to their underweight male counterparts. Kaur et al. (2015) studied on haemoglobin status and body mass index (BMI) in adolescents and concluded that nutritional anemia and under-nutrition prevailed more among girls rather than boys in spite of being literate and having a healthy nutritive diet. The present study also highlights the fact that the prevalence of anemia is comparatively very high among those Muslim and Meitei individuals who are having low BMI. Gupta et al. (2011) studied the young males and females of rural Punjab and mentioned that the prevalence of anemia significantly increases with a decrease in the BMI. Pal et al., (2014) tried to investigate the interrelationship of anemia with BMI among the adult rural population of the state of West Bengal, India and indicated that the adults with low BMI would have a higher risk of anemia compared to normal or overweight adults. Medhi et al. (2006) revealed widespread prevalence of thinness and anemia among the adults of tea garden population of Dibrugarh district of Assam and remarked that health status of the population can be improved through better hygienic practices, environmental sanitation, creating health awareness, nutritional intervention and overall improvement of socio-economic conditions of the population.

Conclusion

The study reveals that there is a significant positive correlation between body mass index and haemoglobin concentration level of the adults in both the communities irrespective their sex but correlation value is more among the Muslims compared to the Meiteis. It is observed that the severity of anemia increases with the decrease in the level or grades of body mass index (nutritional status) in both the communities. Sex wise distribution of different grades of anemia also gives a similar picture. The study also highlights the fact that the prevalence of anemia is significantly high among those Muslim and Meitei individuals who are underweight compared to the non-underweight individuals of both the communities.

References

Basu, D 2006, Nutritional anthropometric assessment of nutrients' intake among the Dimasa tribal adults of North Cachar Hills, Assam. Basu, D, Kulirani, BF, and Ray, BD (eds.) *Agriculture, Food security, Nutrition, and Health in Northeast India* New Delhi, Mittal Publications, pp. 363-381.

Basu, D, Kumar, V and Reddy, BM 2005, Genetic heterogeneity and population structure: A study of northeast India with reference to neighboring populations. Das, RK, and Basu, D (*eds.*) *Northeast India in Perspective-Biology, Social formation, and Contemporary Problems,* New Delhi, Akansha Publishing House, pp38-59.

Bentley, M E and Griffiths, PL 2003, 'The burden of anemia among women in India'. *European Journal of Clinical Nutrition*, Vol. 57, no.1, pp. 52-60.

Bharati, P, Ghosh, R and Gupta, R 2004, 'Socio-economic condition and anemia among the Mahishya population of Southern West Bengal, India'. *Malaysian Journal of Nutrition*, Vol. 10, no. 1, pp. 23-30.

Bhargava, A, Bouis, H E and Scrimshaw, NS 2001, 'Dietary intakes and socio-economic factors are associated with the hemoglobin concentration of Bangladeshi women'. *Journal of Nutrition*, Vol. 131, pp. 758-764.

Bhattacharjee, S, Banerjee, R, Roy, J K, Mandal, S, Biswas, R and Chakraborty, M 2010, 'Undernutrition and anemia in rural adults- A cross-sectional study in rural North Bengal'. *Indian Journal of Preventive and Social Medicine*, Vol. 41, no. 1 and 2, pp. 33-36.

Biswas, M and Baruah, R 2014, 'Maternal anaemia associated with socio-demographic factors among pregnant women of Boko-Bongaon Block Kamrup, Assam'. *Indian Journal of Basic and Applied Medical Research*, Vol. 3 no. 2, pp. 712-721.

Ghosh, R and Bharati, P 2003, 'Haemoglobin status of adult women of two ethnic groups living in a peri-urban area of Kolkata city, India: a micro-level study'. *Asia Pacific Journal of Clinical Nutrition, Vol.* 12, no. 4, pp. 451-459.

Gupta, VK, Maria, AK, Kumar, R, Bahia, JS and Arora, S 2011, 'To study the prevalence of anaemia in young males and females with respect to the age, body mass index (BMI), activity profile and the socio-economic status in rural Punjab'. *Journal of Clinical and Diagnostic Research*, Vol. 5, no. 5, pp. 1020-1026.

Haloi, A and Limbu, DK 2013, 'Nutritional status, fertility and factors associated with anemia: A cross-sectional study among a rural population of Muslim women in Assam, India'. *Anthropologischer Anzeiger*, Vol. 70, no.2, pp. 193-201.

IIPS 2002, *National Family Health Survey (NFHS-2), 1998-99, India, National Family Health Survey-2 (NFHS-2), Anemia among women and children*. Mumbai, International Institute for Population Sciences.

IIPS 2007, *International Institute for Population Sciences and Macro International, National Family Health Survey (NFHS-3), 2005-06: India: Volume I*. Mumbai, International Institute for Population Sciences.

Kaur, M, Singh, A, Bassi, R and Kaur, H 2015, 'Nutritional status and anaemia in medical students of SGRDIMSAR, Amritsar'. *National Journal of Physiology, Pharmacy, and Pharmacology,* Vol. 5, no.1, pp. 35-49.

Khongsdier, R 2001, 'Body mass index of adult males in 12 populations of northeast India'. *Annals of Human Biology*, Vol. 28, no.4, pp. 374-383.

Khongsdier, R, and Basu, D 2004, Nutritional status of the Dimasa of North Cachar, Assam. Bhattacharya, KK, Choudhury, SS and Tyagi, D (eds.) *Nutritional Status of Indian Population: Northeast India* Kolkata, Anthropological Survey of India, pp. 45-58.

Leslie, J 1997, Improving the nutrition of women in the third world. Anderson, P Pelletier, D and Alderman, H (eds.) *Child Growth and Nutrition in the Developing Countries* Delhi, Oxford University Press, pp. 117-138.

Medhi, G K, Hazarika, N C, Shah, B and Mahanta, J 2006, 'Study of health problems and nutritional status of tea garden population of Assam'. *Indian Journal of Medical Sciences*, Vol. 60, no.12, pp. 496-505.

Pal, A, De, S, Sengupta, P, Maity, P and Dhara, PC 2014, 'An investigation on the prevalence of Anaemia in relation to BMI and nutrient intake among the adult rural population of West Bengal, India'. *Epidemiology, Biostatistics, and Public Health*, Vol. 11, no. 2, pp. 1-10.

Ramachandra, SS and Kasthuri, A 2008, 'Anemia in the elderly who resided in a south Indian rural community'. *Indian Journal for the Practicing Doctor*; Vol. 10, no. 5(4): 2.

Saikia, Y 2004, *Fragmented Memories: Struggling to be Tai-Ahom in India.* Durham, N.C., Duke University Press.

Singh, MR and Mangang, K 2012, 'Anaemia and body mass index (BMI) of fisherwomen inhabiting in Karang island of Loktak Lake, Manipur (India)'. *Eurasian of Journal of Anthropology*, Vol. 3, no. 2, pp. 47-53.

Weiner, JS, and Lourie, JA 1981, *Practical human biology*. London, Academic Press.

World Health Organisation 1995, *Physical status: The use and interpretation of anthropometry*, technical report series no. 854. Geneva, World Health Organization.

World Health Organisation 2000, *Obesity: preventing and managing the global epidemic*, report of a WHO consultation, technical report series no. 894. Geneva, World Health Organization.

World Health Organisation 2001, *Iron deficiency anaemia assessment, prevention, and control: A guide for programme managers*. Geneva, World Health Organization.

World Health Organisation 2011, *Haemoglobin concentrations for the diagnosis of anaemia and assessment of severity,* vitamin, and mineral nutrition information System (WHO/NMH/NHD/MNM/11.1). Geneva, World Health Organization.

World Bank 1991, *Gender and poverty in India: A world bank country study*. Washington DC: The World Bank.

http://www.cachar.nic.in accessed on 01-12-2014.

http://cachar.gov.in accessed on 19-06-2015.

http://www.census2011.co.in/district.php (District Census, 2011) accessed on 30-02-2015.

http://familypedia.wikia.com/wiki/Cachar_district("Cachar district," n.d.) accessed on 12-03-2015.

Chapter - 5

Physical Growth Assessment among Indigenous Karbi Children of Assam, India: Comparison of WHO, CDC, and NHANES Growth References

— Nitish Mondal, Harbamon Rongpi, and Kaushik Bose

Introduction

Physical growth assessment can play an important role in nutritional status due to the effect of age, sex, ethnic, geographic, environmental, socio-economic conditions, sedentary behaviour, physical activity, disease and genotype or genetic factors and provides an useful indication of the overall health and nutritional status (Eveleth and Tanner, 1991; Rolland-Cachera, 1993; Rogol et al., 2002; Rogol, 2010; Wells, 2010; Sen and Mondal, 2013; Rengma et al., 2016; Mondal et al., 2017; Sharma and Mondal, 2018). Undernutrition during childhood can affect physical growth potential and increases the risks of morbidity and mortality among children have special nutritional needs because of the extensive rate or tempo of physical growth and development (Nandy et al., 2005; Semproli and Gualdi-Russo, 2007; Olivieri et al., 2008). Developing countries like India continue to have a high prevalence of childhood undernutrition, mortality, and morbidities (Nandy et al., 2005; Bamji et al., 2008; Varadharajan et al., 2013). High prevalence of undernutrition among children is considered to be a major public health problem in rural India (Bose et al., 2007; Mondal and Sen, 2010; Sen and Mondal, 2012; Debnath et al., 2018). Several researchers have opined that rural environment is the important factor which delay the optimal physical growth pattern and body composition in many developing countries including India (Semproli and Gualdi-Russo, 2007; Olivieri et al., 2008; Andrissi et al., 2013; Mondal et al., 2017; Xu and Hang, 2017; Zong et al., 2017; Sharma and Mondal, 2018).

India (given its large population size and widespread poverty and a large proportion of individuals being undernourished and underprivileged), has the highest prevalence of the childhood undernutrition in the world and it has been estimated that more than half of Indian children are undernourished (Svedberg, 2011; Ramachandran, 2014). Significant numbers of studies have reported the poor attainment of physical growth and high prevalence of undernutrition among the different tribal and non-tribal population of India (Khongsdier, 2005; Mondal et al., 2017; Sharma and Mondal, 2018). However, a limited data are available on the anthropometric assessment of physical growth and nutritional status of tribal populations of India (Sen and Mondal, 2013; Singh and Mondal, 2014; Debnath and Bhattacharjee, 2016; Debnath et al., 2017). Body Mass Index (BMI =weight (kg)/height $(m)^2$) a derived anthropometric measure played a significant role in the assessment of physical growth and nutritional status (Sen and Mondal, 2013; Rengma et al., 2015; Mondal et al., 2017; Sharma and Mondal, 2018). Further, several researchers have opined that BMI is not able to desegregate the proportions of body adiposity (fat mass) to lean body mass (fat free mass) in clinical/epidemiological investigations (Wells, 2010; Thibault et al., 2012; Thibault and Pichard, 2012; Sen and Mondal, 2013; Alpizar et al., 2017). Several recent studies have also established the effects of several socio-economic and demographic variables with BMI to assess the nutritional status among children and adults (Subramanian and Smith, 2006; Mondal et al., 2015; Rengma et al., 2015; Debnath et al., 2018).

India comprises of numerous indigenous populations and has an enormous amount of ethnic and genetic diversity (Indian Genome Variation Consortium, 2008). The popularly known Northeast India is a land of innumerable and diverse ethnic population heterogeneity. In this region, there are about 130 tribal groups constituting about one-third of the total number of tribes in India. Given the region's backwardness, the majority of the population (children and adolescents particularly) are nutritionally vulnerable. High prevalence of undernutrition is generally observed to be a serious public health problem of the populations of Northeast India (Khongsdier and Mukherjee, 2003; Agahar-Murugkar, 2005; Medhi et al., 2006; Basu et al., 2013; Singh and Mondal, 2014; Mondal et al., 2015; Rengma et al., 2016; Kramsapi et al., 2018; Sharma and Mondal, 2018). The populations residing in rural environments are found to be more vulnerable and underprivileged in terms of poverty, literacy, nutritional deprivation (undernutrition). In general, poor or inadequate healthcare situations exist in rural India. An extensive literature search reveals that very few studies have been reported on physical growth pattern and body composition assessment among children and adolescents in Northeast, India (Basu et al., 2010; Islam et al., 2014; Singh and Mondal, 2014; Sharma and Mondal, 2018). The assessment physical growth pattern among the nutritionally vulnerable segments of the population bears significant importance in public health in order to identify and/or reduce the risk of nutritional disorders (undernutrition or overweight and obesity). Assessment of physical growth and nutritional status will help the policy makers and social scientists to design appropriate nutritional intervention programmes. Therefore, the objectives of the present study were to assess and compare the physical growth pattern and undernutrition status with international growth references among Karbi children residing in rural areas of Karbi Anglong, Assam.

Material and Methods

Subject and Study Area

The present cross-sectional study was carried out among the 401 school-going children aged 5 to 12 years of an indigenous Karbi population of Karbi Angling district of Assam, Northeast India. The subjects were inhabitants of three villages, namely, Langmili, Rongchingbar, Shirkangnep village of Diphu, Karbi Anglong, Assam. The studied villages are mostly inhabited by the homogeneous Karbi population. The Karbi Anglong district (25°33' N to 26°35' N Latitude and 92°10' E to 93°35'E Longitude) occupies an area of 10,434 km sq. and is the largest District in Assam. According to the census of 2011, Karbi Anglong had a population of 9, 65,280 of which 4, 93,482 and 4, 71,798 are male and female respectively. The literacy rate of Karbi Anglong according to the census of 2011 is 73.52 percent (82.12% male; 64.62% female). Ethnically, Karbi belongs to the Mongoloid ethnic group and linguistically it belongs to the Tibeto-Burman speaking group. To avoid the necessary subject selection bias, all the subjects were free from any physical or limb deformities. The present study was carried out in accordance with the ethical guidelines of the human experiment as laid down in the Helsinki Declaration (Touitou et al., 2004; Portaluppi et al., 2010). The data was collected during the period of November 2016 to March 2017. The household survey method was employed in order to collect the socio-economic and demographic variables. The parents of the children were interviewed for obtaining the relevant data on socio-economic and demographic variables. The study objectives were explained and verbal consent was taken from the parents prior to collect the anthropometric and socio-economic data, demographics and lifestyles data. The children were included using a stratified random sampling method. The surname, physical and cultural features were also utilized to ascertain the ethnicity of the children. The ethnicity of the subjects was also verified using the official documents issued by government authorities. The data of the present investigation were collected employing pre-structured schedules. Children's parents were systematically interviewed for obtaining the relevant data on material risk factors, socio-economic and demographic variables. The minimum number of research participants required for reliably estimating the undernutrition and physical growth pattern was calculated following the standard sample size estimation method (Lwanga and Lemeshow, 1991). In this method, the anticipated population proportion of 50.0 percent, absolute precision of 5.0 percent ($\leq$ 5 %; i.e., the lower margin of error) and confidence interval of 95.0 percent are taken into consideration. Therefore, the minimum number of sample size estimated for the present investigation was 385 individuals. Initially, 425 children in the age group of 5-12 years were identified and approached to participate in the investigation. However, 24 children (14 boys and 10 girls) were excluded due to age and/or exclusion criteria of the present investigation. Hence, the final sample size consisted of 401 children (247 boys; 154 girls) research participants included in the present investigation.

Collection of Socio-economic Data

Socio-economic and demographic data on age, sex, nature of school father's occupation and mothers occupation, number of siblings, house number of living rooms, family type, sanitation facility, electric facility, sources of drinking water. In order to elicit valid responses, ample care was taken while briefing the questions to the respondents at the time of investigation. The socio-economic status (SES) was evaluated using a modified version of the scale of Kuppuswamy (Kumar et al., 2007; Gadhave and Nagarkar, 2015). This scale allows determination of SES based on a score calculated from education, occupation and monthly family income status of the parents who have a combination of the social and economic scale. Based on the above-mentioned scale, the children belonged to lower to middle income in SES groups.

Collection of Anthropometric Data

Anthropometric measures were obtained using standard procedures (Hall et al., 2007). The height was taken with the help of anthropometric rod, the subjects were made to stand erect and looking straight at the

Frankfort Horizontal Plane nearest to 0.1 cm. Weight was taken using a portable weighing machine with minimum clothes nearest to 0.5 kg. The subjects were free from any physical deformities and not suffering from any chronic disease at the time of data collection. A technical error in measurements (TEM) was calculated following the method of Ulijaszek and Kerr (1999) to determine the accuracy of the measurements. The intra-observer TEM were observed to be within the cut-offs value of 0.95 as suggested by Ulijeszek and Kerr (1999). Hence, the measurements recorded in the present study are considered to be reliable and reproducible.

Assessment of Nutritional Status

The assessment of nutritional status in the present study was based on the conventional anthropometric index of thinness (low BMI for age) using the recently proposed international references and cut-offs proposed by Cole et al. (2007). Where the prevalence of thinness was categories into three categories included Grade I (mild thinness), Grade II (moderate thinness) and Grade III (severe thinness). These nutritional categories were similar to the proposed cut-offs of Chronic Energy Deficiency (CED) or undernutrition in adults (WHO, 1995).

Statistical Analysis

The data were statistically analyzed using relevant tests using the Statistical Package for Social Science (SPSS, Version 17.0). One-way analysis of variance (ANOVA) was used to assess the age-specific mean differences in anthropometric variables. Independent sample t-test analysis was done to assess the sex-specific mean differences in anthropometric variables. The chi-square analysis was used to access the difference in different grades of nutritional status between the sexes of thinness. The LMS model is a concept of an age varying adjustment for skewness based on the Box-Cox transformation. The method converts the measurements for a subject of known age and sex to evaluate percentile and standard deviation score or Z-score (Cole and Green, 1992; Cole et al., 1998). The LMS method software computer program fits smooth percentile curves to the reference data using the LMS method as described by Cole and Green (1992). The LMS Chartmaker Light programme (version 2.54) was used to derive smooth percentile growth curves (Pan and Cole, 2011). The age and sex-specific height, weight and BMI percentile curves with 3rd, 10th, 15th, 25th, 50th, 75th, 90th, and 97th smoothed percentile lines were plotted separately. A p-value of less than 0.05 was considered being statistically significant.

Results

Descriptive Statistics of Height, Weight, and BMI among Karbi Boys

The age-sex specific subject distributions and mean and standard deviation (± SD) of height, weight, and BMI among Karbi boys are presented in Table 5.1. The overall mean and standard deviation of height, weight, and BMI are 123.07±8.24 cm, 23.43±4.92 kg, and 15.26±1.41 kg/m^2, respectively. The age-specific mean of height, weight, and BMI increased with age among boys. The age-specific mean and SD of height was observed to be 108.23±0.94 cm, 113.57±2.34 cm, 118.23±1.03 cm, 122.80±1.44 cm, 124.88±0.58 cm, 129.14±2.27 cm, 134.70±0.86 cm, 138.67±1.59 cm among the age of order 5 to 12 years of Karbi boys, respectively. The age-specific mean and SD of weight was observed to be 15.85±0.74 kg, 17.83±0.80 kg, 19.99±1.32 kg, 22.69±0.79 kg, 24.81±0.58 kg, 26.77±1.80 kg, 30.27±0.82 kg, 33.51±1.69 kg among the age of 5 to 12 years of Karbi boys, respectively. The age-specific mean and SD of BMI was observed to be higher at age of 12 years (17.43±0.98 kg/m^2) and lower at age of 5 years (13.53±0.69 kg/m^2). Using ANOVA, age-specific mean differences was observed to be statistically significant (Table 5.1) in height (F=943.37; d.f., 1, 7, 246), weight (F=586.85; d.f., 1, 7, 246) and BMI (F=63.73; d.f., 1, 7, 246).

Descriptive Analysis of Height, Weight, and BMI among Karbi Girls

Age-sex specific subject distribution, the mean and standard deviation of height, weight, and BMI among Karbi girls have presented in Table 5.1. The overall mean and standard deviation (±SD) of height, weight, and BMI were 123.05±8.90 cm, 23.13±4.77 kg, 15.09±1.25 kg/m^{2,} respectively. The specific mean of height, weight, and BMI increases with age among Karbi girls. The age-specific mean and SD of height was observed to be 108.08±1.16cm, 113.02±0.60 cm, 118.28±1.08 cm, 123.66±1.25 cm, 125.68±0.63 cm, 130.46±3.48 cm, 134.29±5.73 cm, 139.44±1.91 cm among the age of 5 to 12 years Karbi girls. Similarly, the age-specific mean and SD of weight was observed to be 16.25±1.10 kg, 17.99±1.05 kg, 20.28±1.01 kg, 22.46±0.50 kg, 24.42±0.66 kg, 26.56±0.78 kg, 30.13±1.73 kg, 33.20±0.73 kg among the Karbi girls of age from 5 to 12 year. The age-specific mean and SD of BMI was observed to be higher at age 12 years (17.08±0.57 kg/m^2) and lower at age of 5 years (13.91±1.05 kg/m^2). The age-specific mean differences was observed to be statistically significant

(Table 5.1) in height (F=354.20; d.f., 1, 7, 153), weight (F=586.85; d.f., 1, 7, 153) and BMI (F=63.73; d.f., 1, 7, 153) using ANOVA.

The Sex-specific Trend in Anthropometric Measurements among Karbi Children

The overall sex-specific mean values were observed to be higher among boys than the girls in height, weight, and BMI. The sex-specific differences in ages were not shown any general trends but boys were found to be heavier than the girls in respect to weight and BMI in higher ages, but weight and BMI were observed to be higher among girls than boys in early ages 5-7 years. Sex-specific mean values of height were observed to higher in 5-6 years and 11-12 years among boys, but mean values were observed to higher in 7-11 years in girls. Using independent sample t-test, sex-specific mean differences in anthropometric measurements were observed to be statistically not significant in height (t-value=0.068, d.f., 399), weight (t-value=0.614, d.f., 399) and BMI (t-value=1.232, d.f., 399) among boys and girls ($p>0.01$).The age-sex specific smooth percentile curves with 3rd, 10th, 15th, 25th, 50th, 75th, 90 and 97th smoothed percentile derived using the LMS procedure for height, weight and BMI for further evaluation of physical growth and nutritional among the children is shown in Figure 5.1.

Comparison of Anthropometric Measures with the WHO Growth Reference

The comparison of age-sex specific mean anthropometric variables with the WHO (2007) growth references showed that the majority of the children reflecting inappropriate poor physical growth attainments and nutritional status than their normal counterparts (Figure 5.2). The age-specific comparison showed that mean height values were below 25th percentile among boys (in 8-11 years) and girls (in 9-11 years). The age-specific mean value was found below the 5th percentile in 12 years in both sexes. The age-specific mean weight values were found to be below the 25th percentile of the reference among boys and girls aged 5-10 years, except in 7 years (in girls). Age-specific mean BMI values were observed to be below the 25th percentile and 50th percentile aged 5-8 years and 9-12 years among boys and girls, respectively.

Comparison of Anthropometric Measures with the CDC Growth Reference

The comparison of age-sex specific mean anthropometric variables with the CDC-2000 (Kuczmarski et al., 2002) growth references showed that the majority of the children reflecting poor physical growth attainments and nutrition than their normal counterparts (Figure 5.3). The age-specific comparison showed that mean height values were below 50th percentile among 5-7 years (in boys) and 5-6 years (in girls), while age-specific mean values were found below 25th percentile among 8-12 years (in boys) and 7-11 years (in girls). However, the age-specific mean value was found below the 5th percentile in 12 years in girls only. The age-specific mean weight values were found to be below the 25th percentile of the CDC reference among boys and girls aged 5-12 years. Age-specific mean BMI values were observed to be below the 25th percentile and 50th percentile aged 7-8 years and 9-12 years respectively, but mean values are exceptionally low in 5-6 years (below 5th percentile) among boys. Similarly, age-specific mean BMI values were observed to be below the 25th percentile and 50th percentile aged 5-9 years and 10-12 years among girls, respectively.

Comparison of Anthropometric Measures with the NHANES-III Growth Reference

The comparison of age-sex specific mean anthropometric variables with the NHANES (Frisancho, 2008) growth references showed that the majority of the children reflecting poor physical growth attainments and nutritional status than their normal counterparts (Figure 5.4). The age-specific comparison showed that mean height values were below 25th percentile among 5-8 years and 11 years (in boys) and 5-9 years (in girls), while age-specific mean values were found below 5th percentile among 9 years, 10 years and 12 years (in boys) and 10-12 years (in girls). The age-specific mean weight values were found to be below the 25th percentile of the NHANES reference among boys and girls aged 5-12 years. Age-specific mean BMI values were observed to be below the 25th percentile and 50th percentile of NHANES aged 5-7 years and 9-12 years respectively among boys and girls, respectively.

Prevalence of Thinness among the Karbi Children

Sex-specific prevalence of different grades of thinness among the Karbi children is presented in Table 5.2. The overall, severe, moderate and mild thinness prevalence was observed to be 19.45, 1.50, 4.24, and 13.72 percent, respectively. The sex-specific overall prevalence of mild thinness (12.96% vs. 14.94%) was found to be higher among girls, whereas moderate thinness (5.26% vs. 2.60%) and severe thinness (2.43% vs. 0.00%) was observed to be higher among boys. The age-specific prevalence of thinness was observed to be higher among 6 years and 5 years among boys and girls, respectively.

Similarly, the lower prevalence of thinness was among the 8 and 12 years (1.96%) and 8 years (7.96%) among boys and girls respectively. However, the age-specific difference in the prevalence of thinness between the boys and girls was observed to be statistically non-significant using chi-square analysis.

Table 5.1: Age-specific descriptive statistics of anthropometric variables among rural Karbi children

Age	*Height (cm)*		*Weight (kg)*		*BMI (kg/m2)*	
	Boys	*Girls*	*Boys*	*Girls*	*Boys*	*Girls*
5 years	108.23±0.94	108.09±1.16	15.85±0.75	16.25±1.11	13.53±0.70	13.92±1.05
6 years	113.57±2.34	113.02±0.60	17.83±0.80	18.00±1.05	13.84±0.70	14.09±0.85
7 years	118.23±1.03	118.28±1.08	20.00±1.33	20.28±1.01	14.30±0.93	14.50±0.78
8 years	122.80±1.44	123.67±1.25	22.70±0.80	22.47±0.51	15.06±0.64	14.70±0.47
9 years	124.88±0.58	125.68±0.63	24.81±0.58	24.42±0.67	15.91±0.38	15.46±0.47
10 years	129.14±2.27	130.47±3.48	26.78±1.80	26.57±0.78	16.08±1.25	15.64±0.98
11 years	134.70±0.86	134.30±5.74	30.27±0.83	30.14±1.74	16.69±0.48	16.78±1.68
12 years	138.68±1.59	139.45±1.91	33.51±1.69	33.21±0.73	17.43±0.98	17.09±0.58
Total	123.07±8.24	123.02±8.91	23.44±4.92	23.13±4.77	15.30±1.41	15.10±1.25
F-value	943.37**	354.20**	586.85**	548.56**	63.73**	27.85**

**p<0.01

Table 5.2: Prevalence of thinness among rural Karbi children

Sex	*Prevalence of different grades of thinness*					*Total*
	Severe (Grade-III)	*Moderate (Grade-II)*	*Mild (Grade-I)*	*Total*	*Normal*	
Boys	6 (2.43)	13 (5.26)	32 (12.96)	51 (20.65)	145 (58.70)	247 (100.00)
Girls	0 (0.00)	4 (2.60)	23 (14.94)	27 (17.53)	100 (64.94)	154 (100.00)
Total	6 (1.50)	17 (4.24)	55 (13.72)	78 (19.45)	296 (73.82)	401 (100.00)

Values are in parenthesis indicates the percentage

Discussion

The assessment of physical growth and nutritional status enable to identify the potential risks of undernutrition, overweight and obesity during childhood and adolescence. Therefore, the assessment of the undernutrition among various nutritionally vulnerable segment of populations bears a great significance in developing countries such as India, where the vast proportion of the populations are undernourished and underprivileged (Nandy et al., 2005; Bose et al., 2007; Mondal and Sen, 2010; Sen and Mondal, 2012; Singh and Mondal, 2013; Debnath et al., 2017; Sharma and Mondal, 2018). Such assessment studies will definitely be able to identify the risks of physical growth, nutritional problem, and subsequent health issues in individual and/or populations. There is substantial evidence of a decline

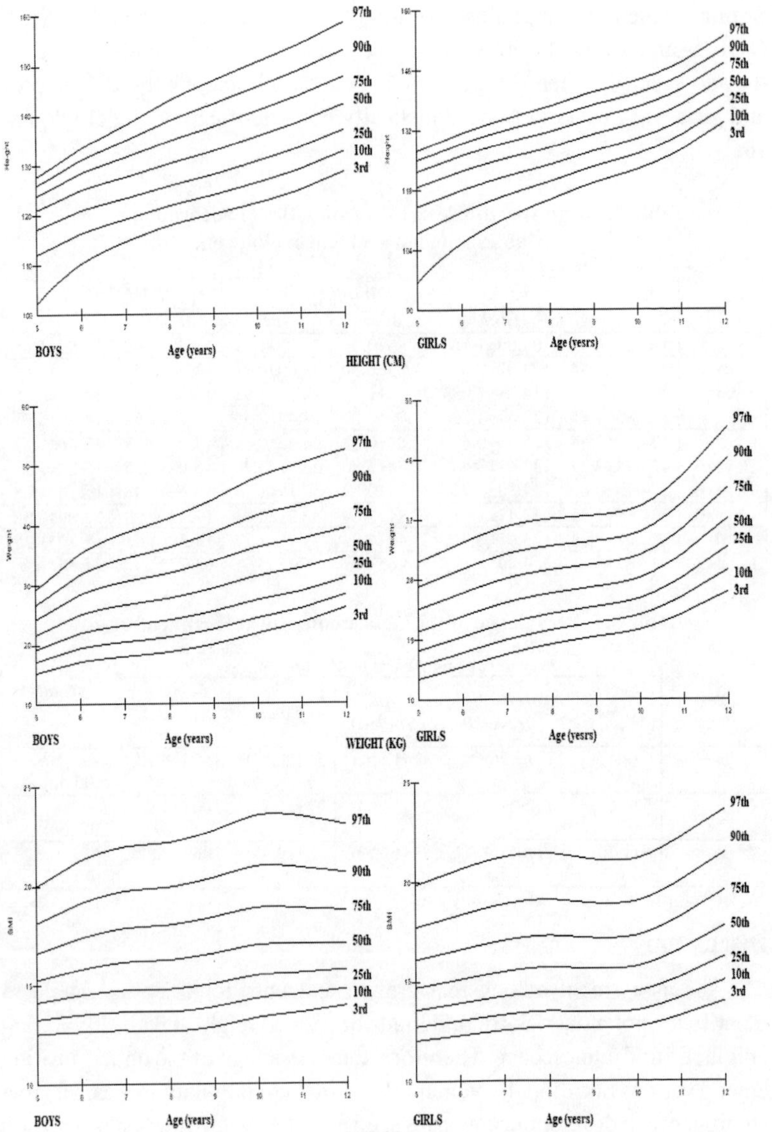

Figure 5.1: Age-sex specific LMS references curve of height, weight, and BMI among rural Karbi children.

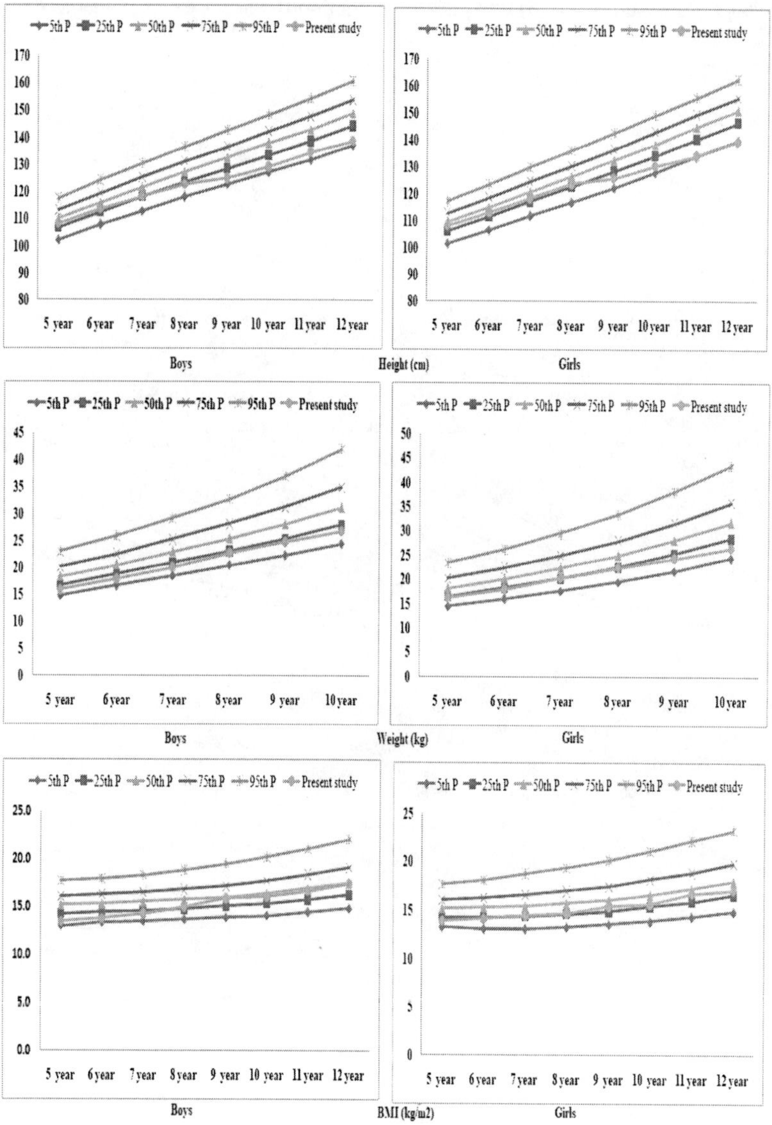

Figure 5.2: Age-sex specific mean comparison of height, weight, and BMI with WHO growth references among rural Karbi children.

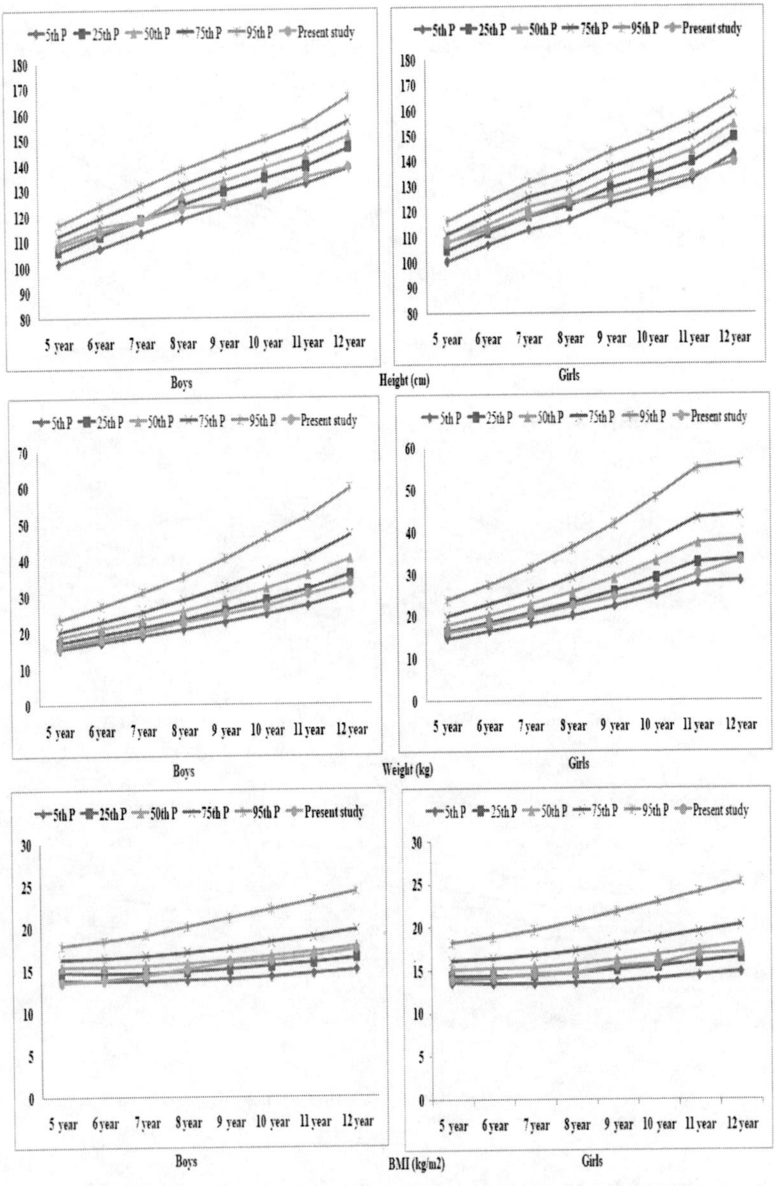

Figure 5.3: Age-sex specific mean comparison of height, weight, and BMI with CDC-2000 growth references among rural Karbi children.

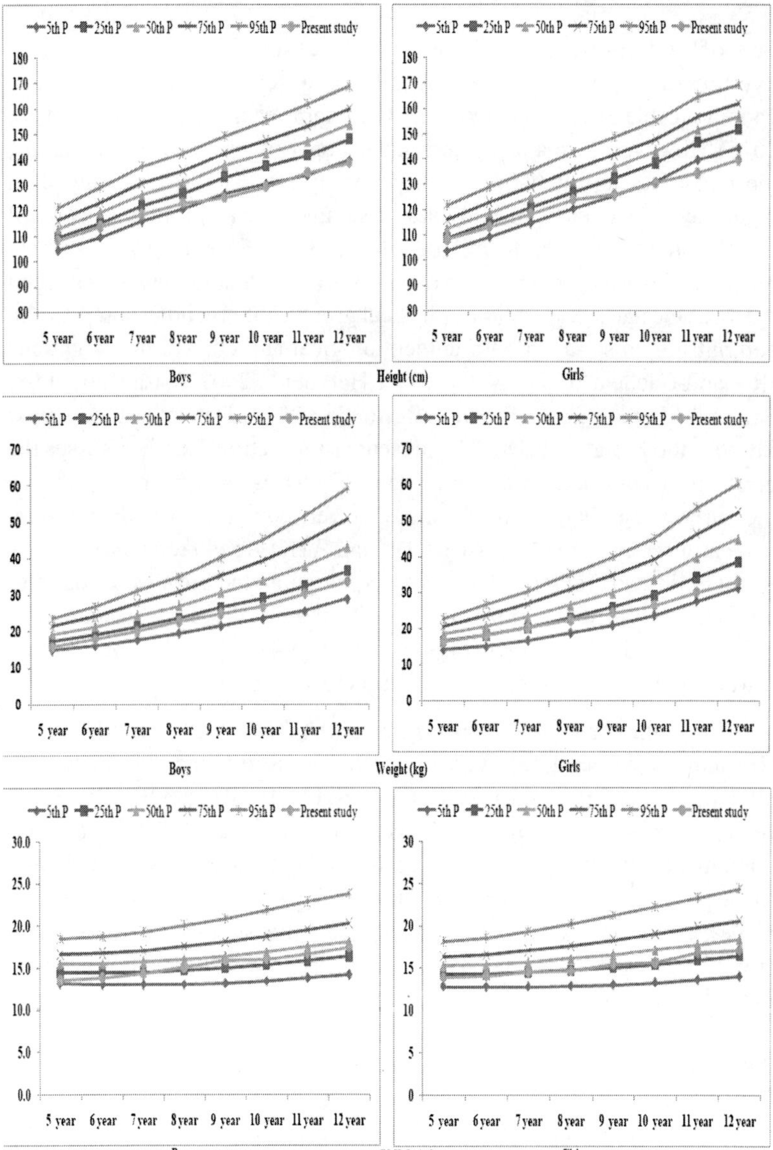

Figure 5. 4: Age-sex specific mean comparison of height, weight, and BMI with NHANES-III growth references among rural Karbi children.

rate of undernutrition prevalence with a simultaneous increase in overnutrition (overweight or obesity) and several preventable non-communicable diseases (Popkin, 2006; Rengma et al., 2015; Mondal et al., 2015). However, the major concern and national policy response for tackling the undernutrition in India is still primarily restricted to undernutrition diagnosed on the basis of body size assessment or anthropometry (Kapil and Sachdev, 2012; Vardharajan et al., 2013). Anthropometry is a single most non-invasive, inexpensive and easy-to-use technique is widely used to assess the age/sex variations, physical growth pattern, body composition and nutritional status in both epidemiological and clinical investigations (Rolland-Cachera, 1993; WHO, 1995; Hall et al., 2007; Frisancho, 2008; Sen and Mondal, 2013; Singh and Mondal, 2014; Debnath et al., 2017; Sharma and Mondal, 2018). The present cross-sectional study assesses the physical growth pattern and nutritional status using the anthropometric measurements of height, weight and BMI were compared with the different percentile value of NHANES-III, CDC and WHO growth references among Karbi children (Figures 5.2 and 5.4). Apparently, the boys were found to be undernourished (<25th percentile) with the decreases advancement (in height) towards higher ages (Figures 5.2 and 5.4), especially approaching towards puberty or adolescents (10 years to 11 years).

The comparison with growth references (WHO, CDC and NHANES-III) showed that children were found to be more vulnerable to growth attainment (height; <5th percentile) with NHANES-III than WHO or CDC reference in both sexes aged 10-12 years (Figures 5.2 and 5.4). Such poor attainment of physical growth especially in higher ages may be attributed to the poor access to food or inadequate intake or nutritional deprivation in respect to physical requirements, socio-economic status, and environment (rural). Several researchers have reported the poor attainment of physical growth and undernutrition in comparison with international growth references among Indian tribal and non-tribal children (Reddy and Rao, 2000; de Onis et al., 2001; Medhi et al., 2006; Singh and Mondal, 2014; Debnath et al., 2017; Sharma and Mondal, 2018). Deficiency of the required nutrients may lead to the prevalence of chronic undernutrition and/or inadequate physical growth attainment and body composition status, which is identified with reduced lean body mass, low muscular strength and working capacity (WHO, 1995; Medhi et al., 2006; Mondal and Sen, 2010; Sharma and Mondal, 2018). The poor attainment of physical growth and body composition leads to several physical manifestations including delays in menarche attainments and poor reproductive outcomes or low birth weight (WHO, 1995) and also resulting the physical growth retardation and manifest undernutrition in population (Medhi et al., 2006; Semproli and Gualdi-Russo, 2007; Olivieri et al., 2008; Singh and Mondal, 2014; Debnath et al., 2017; Sharma and Mondal, 2018). Several research studies have also confirmed that physical growth retardation or low height-for-age (stunting) in early life results in an adaptation in energy metabolism that speed up body fat deposition or adiposity and increases the relative risk of children becoming overweight/obese and likelihood of suffering from the metabolic disorders later in life (Sawaya et al., 2004; Lee et al., 2015).

Recently, the WHO has published child growth standards for attained weight and height to replace the previously recommended growth reference data of NCHS/WHO (WHO, 1983). These new standards were based on breastfed infants and appropriately fed children of different ethnic origins, raised in optimal conditions and measured in a standardized manner (de Onis et al., 2007). These reference data were proposed so as to identify the individual at risk, assess response to intervention and facilitate international comparisons (de Onis et al., 2001). However, the NCHS reference data (WHO, 1995) has been found to be consistent with these currently advocated international growth references for children (WHO, 1995; de Onis and Habicht, 1996). Given these issues both the CDC-2000 (Kuczmarski et al., 2002), NHANES-III (Frisancho, 2008) and WHO (2007) reference data have been utilized to assess the physical growth pattern among indigenous Karbi children. Several studies have evaluated the comparative growth and nutritional status among children and adolescents using different reference populations where most of the studies have utilized NCHS or NHANES, CDC and WHO reference population (de Onis et al., 2007; Hanicar et al., 2009; Silveira et al., 2011; Oliveira et al., 2013). The age-specific mean comparison in physical growth attainments was found to be also lower than NHANES-III (Frisancho, 2008), CDC-2000 (Kuczmarski et al., 2002) and WHO (2007) growth references among Karbi children (Figures 5.2 and 5.4). The comparative studies showed similar results that the poor attainment of physical growth prevalence of undernutrition is found higher in NCHS or NHANES, CDC than WHO growth reference (Rousham et al., 2011; Silveira et al., 2011). The WHO standards provide a relatively better tool to monitor the rapid and changing rate of physical growth pattern and assess the undernutrition in early infancy and childhood. The adoption of WHO (2007) reference will have important implications for child health or nutritional status with respect to the assessment of lactation performance and the adequacy of infant feeding (de Onis et al., 2006). Saha et al. (2009) confirm that use of the NCHS reference misidentifies undernutrition and the timing of growth faltering in infants and young children, which was a key rationale for constructing the new WHO standards. Several research studies have advocated the use of WHO (2007) proposed reference population is consisted to assess the appropriate level of undernutrition or highest possible number of undernutrition and nutritional risk factor groups for the early intervention in infancy and childhood or who should get the early nutritional interferences both in clinical and epidemiological investigations (de Onis et al., 2007; Schwarz et al., 2008; Rousham et al., 2011; Silveira et al., 2011).

In the present study, the assessment of thinness among the Karbi children of Karbi Anglong district in Northeast India was undertaken using a newly proposed international cut-off of Cole et al. (2007). This new cut-off point was suggested to encourage direct comparison of trends in child and adolescence thinness prevalence worldwide. This cut-off provide classification of thinness for a public health proposes prior to this report there were no thinness cut-off points for this age group (2 years to 18 years) except the National Centre of Health Statistics (Must et al., 1991) and NHANES reference population for children and adolescents (WHO, 1995). It must be mentioned here that these cut-off points were derived from multicentre data from the United States, Great Britain, Hong Kong, and the Netherlands, including data from Brazil (developing country). Several studies have already reported the magnitude of thinness prevalence among children and adolescents in Indian population (Biswas et al., 2009; Chakraborty and Bose, 2009; Bisai and Manna, 2010; Mondal and Sen, 2010; Das and Bose, 2011; Singh and Mondal, 2013; Sharma and Mondal, 2014; Tigga et al., 2015), utilising these newly proposed thinness cut-off points of Cole et al. (2007). Hence, cut-off points are valid for the assessment of thinness among Karbi children of Assam, Northeast India. It is also apparent from this study, that there is high prevalence (19.45%) of overall thinness, observed among the Karbi children. In the present study, the boys (20.65%) were found to be more sufferer (Table 5.2) than their girl (17.53%) counterparts. Several studies have reported that boys were more sufferer than girls in thinness prevalence among children and adolescents in India (Chakraborty and Bose, 2009; Bisai and Manna, 2010; Bisai et al., 2010; Mondal and Sen, 2010; Mondal et al., 2015). It is also evident that the prevalence of overall thinness was found to be higher among Bengali (62.26%; Chakraborty and Bose, 2009), rural children of Darjeeling (63.40%; Mondal and Sen, 2010), Kora-Modi tribe (67.23%; Bisai et al., 2010), Santal tribe (56.40%; Das and Bose, 2011) and found lower among the Nepali children (45.15%; Das and Banik, 2001). Mondal and Sen (2010) have also reported the high prevalence of thinness among rural adolescent

(49.10%) of Darjeeling district, West Bengal. Moreover, the assessment and comparison of undernutrition status are considered to be essential for the improvement of the overall health and development or target specific implementation of a nutritional intervention programme in population.

Conclusion

The results of this study indicate that poor physical growth attainment and undernutrition are still an important problem among Karbi children aged 5 to 12 years of Karbi Anglong district, Assam, Northeast India. There were significant sex-related physical growth attainment differences observed between boys and girls, but boys were observed to be more affected by undernutrition than girls in most cases. However, with the advancement of age towards puberty, the height and weight fall below the normal standards among children gradually fall as they approach adolescence and/or due to inadequate nutrition during early adolescence. The appropriate nutritional intervention programmes are necessary for the form of nutritional intervention among the target population to ameliorate the undernutrition problem. The interventional should be introduced in terms of a properly balanced diet and protective nutrient to reduce the mortality, morbidity and ameliorate the nutritional status.

Acknowledgement

The authors gratefully acknowledge the valuable help and cooperation of school authorities, parents, research participants, and local community and village leader during the fieldwork. The extended help and cooperation of the Department of Anthropology, Assam University (Diphu Campus) is also acknowledged.

References

Agrahar Marugkar, D 2005, 'Nutritional status of Khasi school going girls in Meghalaya'. *Nutrition*, Vol. 21, pp. 45-31.

Alpizar, M, Peschard, VG, Escalante-Araiza, F, Altamirano- Bustamante, NF, Murata, C, Arenas-Pérez, R and Rodriguez-Ayala, E 2017, Smoothed body composition percentiles curves for Mexican children aged 6 to 12 years. Children (Basel), 4: E112. pii: E112. doi: 10.3390/children 4120112.

Andrissi, L, Mottini, G, Sebastiani, V, Boldrini, L and Giuliani, A 2013, 'Dietary habits and growth: An urban/rural comparison in the Andean region of Apurimac, Peru'. *Ann, I Super Sanita,* Vol. 49, pp. 340-346.

Bamji, MS 2003. 'Early nutrition and health Indian perspective'. *Current Science,* Vol. 85, pp. 1137-1142.

Bamji, MS, Murthy, PVVS, Williams, L and Vardhana, Rao MVV 2008, 'Maternal nutrition status and practices and perinatal neonatal mortality in rural Andhra Pradesh India'. *Indian Journal of Medical Research,* Vol. 127, pp. 44-51.

Basu, D, Banerjee, I, Sun, D, Bartwal, MS and Devi, RK 2013, 'Cross-sectional reference values for BMI among Khasi tribal adolescents of Meghalaya, India'. *Anthropologischer Anzeiger,* Vol. 70, pp. 179-191.

Basu, D, Sun, D, Banerjee, I, Singh, YM, Kalita, JG and Rao, VR 2010, 'Cross-sectional reference values of upper arm anthropometry of the Khasi tribal adolescents of Meghalaya, India'. *Asia Pacific Journal of Clinical Nutrition*, Vol. 19, pp. 283-288.

Bisai, M, and Manna, I 2010, 'Prevalence of thinness among urban for preschool children in West Bengal, India'. *Sudanese Journal of Public Health*, Vol. 4, pp. 193-198.

Bisai, S, Ghosh, T, De, GK and Bose, K 2010, 'Very high prevalence of thinness among Kora Mudi tribal children of Paschim Medinipur District of West Bengal, India'. *Electronic Journal of Biology,*, Vol. 3, pp. 43-49.

Biswas, S, Bose K, Mukhopadhyay, A and Bhadra, M 2009, 'Prevalence of undernutrition among pre-school children of Chapra, Nadia District, West Bengal, India, measured by the composite index of anthropometric failure (CIAF)'. *Anthropologischer Anzeiger,* Vol. 67, pp. 269-279.

Black, RE, Morris, SS and Bryce, J 2003, Where and why are 10 million children dying every year? *Lancet*, Vol. 361, pp. 2226-2234.

Bose, K, Biswas, S, Ganguly, S, Khatun, A, Mukhopadhyay, A and Bhadra, M 2007, 'Stunting underweight and wasting among Integrated Child Development Services (ICDS) scheme for children aged 3-5 years of Chapra, Nadia district, West Bengal, India'. *Matern Child Nutrition*, Vol. 3, pp. 216-221.

Chakrabarty, S and Bharati, P 2008, 'Physical growth and nutritional status of the Shabar tribal adolescents of Orissa, India: Across-sectional study'. *Malaysian Journal of Nutrition,* Vol. 14, pp. 101-112.

Chakraborty, R, and Bose, K 2009, 'Very high prevalence of thinness using new international body mass index cut-off points among 5 to 10 years old school children of Nandigram, West Bengal, India'. *Journal of Research in Medical Science*, Vol. 14, pp. 129-133.

Cole, TJ, Flega, KM, Nicholls, D and Jackson, AA. 2007, 'Body mass index cut-off to define thinness in children and adolescents: International Survey'. *British Medical Journal*, Vol. 33, no.51, pp. 94-201.

Cole, TJ, Freeman, JV and Preece, MA 1998, 'British 1990 growth reference centiles for weight, height, body mass index and head circumference fitted by maximum penalized likelihood'. *Statistics in Medicine,* Vol. 17, pp. 407-429.

Cole, TJ, and Green, PJ 1992, 'Smoothing reference centile curves: the LMS method and penalized likelihood'. *Statistics in Medicine,* Vol. 11, pp. 1305-1319.

Dapi, LN, Janlert, U, Nouedoui, C, Stenlund, H and Haglin, L 2009, 'Socio-economic and gender differences in adolescents' nutritional status in urban Cameroon, Africa'. *Nutrition Research*, Vol. 29, pp. 313-319.

Das, S and Banik, SD 2001, 'Prevalence of thinness among Nepali speaking preschool children of Darjeeling, using body mass index cut-off points'. *Italian Journal of Public Health,* Vol. 8, pp. 241-246.

Das, S, and Bose, K 2011, 'Prevalence of thinness among Santal preschool children using new body mass index Cut-off points'. *Journal of Anthropological Research,* article ID 602782.

De Onis, M, Dasgupta, P, Saha, S, Sengupta, D and Blossner, M 2001, 'The national center for health statistics reference and the growth of Indian adolescent boys'. *American Journal of Clinical Nutrition,* Vol. 74, pp. 248-253.

De Onis, M, Garza, C, Onyango, AW and Borghi, E 2007, 'Comparison of the WHO child growth standards and the CDC 2000 growth charts'. *Journal of Nutrition,* Vol. 137, pp. 144-148.

De Onis, M and Habicht, JP 1996, 'Anthropometric reference data for international use: recommendations from a world health organization expert committee'. *American Journal of Clinical Nutrition,* Vol. 64, pp. 650-658.

De Onis, M, Onyango, AW, Borghi, E, Garza, C, and Yang, H 2006, 'WHO Multicentre Growth Reference Study Group. Comparison of the world health organization (WHO) child growth standards and the national center for health statistics/WHO international growth reference: implications for child health programmes'. *Public Health Nutrition*, Vol. 9, pp. 942-947.

De Onis, M, Onyango, AW, Borghi, E, Siyam, A, Nishida C, and Siekmann, J 2007, 'Development of a WHO growth reference for school-aged children and adolescents'. *Bulletin of World Health Organisation*, Vol. 85, pp. 660-667.

Debnath, A and Bhattacharjee, N 2016, 'Understanding malnutrition of Tribal children in India: The role of women's empowerment'. *Ecology and Food Nutrition*, Vol. 55, pp. 508-527.

Debnath, S, Mondal, N and Sen J 2018, 'Socio-economic and demographic correlates of stunting and thinness among rural school-going children (aged 5-12 Years) of North Bengal, Eastern India'. *Journal of Life Science*, Vol. 10, pp. 14-31.

Debnath, S, Mondal N and Sen J 2017, 'Use of upper arm anthropometry, upper arm muscle area-by-height (UAMAH) and mid-upper-arm-circumference (MUAC)-for-height as indicators of body composition and nutritional status among children'. *Anthropological Review,* Vol. 80, pp. 85-102.

Eveleth, PB, and Tanner, JM 1991, *Worldwide variation in human growth*. 2nd Edition. Cambridge: Cambridge University Press.

Frisancho, AR 2008, *Anthropometric standards: an interactive nutritional reference of body size and body composition for children and adults*. University of Michigan Press.

Gadhave, S and Nagarkar A 2015, 'Kuppuswamy scale for measuring socio-economic status: revised monthly income figures for 2015'. *Indian Journal of Pediatrics*, Vol. 82, pp. 1175-1176.

Hall, JG, Allanson, JE, Gripp, KW, and Slovotinek, AM 2007, *Handbook of physical measurements*. New York: Oxford University Press.

Hanicar, B, Mandic, Z and Pavic R 2009, 'Exclusive breastfeeding and growth in Croatian infants—comparison to the WHO child growth standards and to the NCHS growth references'. *Collegium Antropologicum,* Vol. 33, pp. 735-741.

Harris, EF and Smith, RN 2009, 'Accounting for measurement error: A critical but often overlooked process'. *Archives of Oral Biol*ogy, Vol. 54, pp. S107-S117.

Indian Genome Variation Consortium 2008, 'Genetic landscape of the people of India: A canvas for disease gene exploration'. *Journal of Genet*ics, Vol. 87, pp. 3-20.

Islam, S, Mahanta, TG, Sarma, R and Hiranya, S 2014, 'Nutritional status of under 5 children belonging to tribal population living in riverine (char) areas of Dibrugarh District, Assam'. *Indian Journal Community Medicine,* Vol. 39, pp. 169-174.

Kapil, U and Sachdev, HP 2012, 'Urgent need to orient public response to rapid nutrition transition.' *Indian Journal Community Med*icine, Vol. 37, pp. 207-210.

Khongsdier, R and Mukherjee, N 2003, 'Growth and nutritional status of Khasi boys in Northeast India relating to exogamous marriages and socio-economic classes'. *American Journal of Physical Anthropology*, Vol. 122, pp. 162-170.

Khongsdier, R 2001, 'Body mass index of adult males in populations of notice India'. *Annals of Human Biology*, Vol. 28, pp. 374-383.

Khongsdier, R 2005, 'BMI and morbidity in relation to body composition: a cross-sectional study of a rural community in North-East India'. *British Journal of Nutrition*, Vol. 93, pp. 101-107.

Kramsapi, R, Singh, N Kh and Mondal, N 2018, 'Composite index of anthropometric failure (CIAF) among preschool (2-5 years) tribal children of Assam (India)'. *Human Biology Review*, Vol. 7, pp. 1-18.

Kuczmarski, RJ, Ogden, CL, Guo, SS, Grummer-Strawn, LM, Flegal, KM, Mei, Z, Wei, R, Curtin, LR, Roche, AF and Johnson CL 2002, 'CDC growth charts for the United States: methods and development'. *Vital Health Statistics*, Vol. 11, pp. 1-190.

Kumar, N, Shekhar, C, Kumar, P, and Kundu, AS 2007, 'Kuppuswamy's socio-economic status scale-updating for 2007'. *Indian Journal of Pediatrics*, Vol. 74, pp. 1131-1132.

Lee, SK, Nam, SY and Hoffman, DJ 2015, 'Growth retardation at the early life and metabolic adaptation among North Korean children'. *Journal of Developmental Origins of Health and Dis*ease, Vol. 6, pp. 291-298.

Lwanga, SK, and Lemeshow, S 1991, *Sample size determination in health studies: A practical manual*. Geneva: world health organization.

Medhi, GK, Hazarika, NC and Mahanta, J 2007, 'Nutritional status of adolescents among tea garden workers'. *Indian Journal of Pediatrics* Vol. 74, pp. 343-347.

Medhi, GK, Hazarika, NC, Shah, B and Mahanta J 2006, 'Study of health problems and nutritional status of the garden populations of Assam, India'. *Journal of Medical Science,* Vol. 60, pp. 496-505.

Mondal, N and Sen, J 2010, 'Prevalence of undernutrition among children 5 to 12 years belonging to three communities residing in a similar habit in North Bengal, India'. *Annals of Human Biology*, Vol. 37, pp. 198-216.

Mondal, N and Sen, J 2010, 'Thinness is a measure underlying problem among Indian children'. *Journal of Tropical Pediatriatics,* Vol. 56, pp. 456-458.

Mondal, N, Basumatary, B, Kropi, J and Bose, K 2015, 'Prevalence of the double burden of malnutrition among urban school going Bodo children aged 5-11 years of Assam, Northeast India'. *Epidemiology, Biostatistics and Public Health*, Vol. 12, pp. 1-11.

Mondal, N, Ghosh, P and Sen, J 2017, 'Physical growth pattern and body composition of Rajbanshi adolescent boys of Eastern India'. *Human Biology Review*, Vol. 6, pp. 263- 283.

Must, A, Dallal, GE, and Dietz, WH 1991, 'Reference data for obesity: 85th and 95th percentiles of body mass index (wt/ht^2) and triceps skinfold thickness'. *American Journal of Clinical Nutrition,* Vol. 53, pp. 839–846.

Nandy, S, Irving, M, Gordon, D, Subramanian, SV and Smith, GD 2005, 'Poverty child undernutrition and morbidity: New evidence from India'. *Bulletin of World Health Organisation,* Vol. 83, pp. 10-216.

Olivieri, F, Semproli, S, Pettener, D and Toselli, S 2008, 'Growth and malnutrition of rural Zimbabwean children (6-17 years of age)'. *American Journal of Physical Anthropology*, Vol. 136, pp. 214-222.

Oliveira, GJ, Barbiero, SM, Cesa, CC and Pellanda, LC 2013, 'Comparison of NCHS, CDC, and WHO curves in children with cardiovascular risk'. *Revista da Associacao Medica Brasileira,* Vol. 59, pp. 375-380.

Pan, H and Cole, TJ 2011, *'LMS Chartmaker, a program to construct growth references using the LMS method'*. Version 2.54. http://www.healthforallchildren.co.uk/.

Popkin, BM 2006, 'Global nutrition dynamics: The world is shifting rapidly towards that linked with non-communicable diseases'. *American Journal of Clinical Nutrition*, Vol. 84, pp. 289-298.

Portaluppi, F, Smolensky, MH and Touitou, Y 2010, 'Ethics and methods for biological rhythm research on animals and human beings'. *Chronobiology International,* Vol. 27, pp. 1911-1929.

Ramachandran, N 2014, *Persisting undernutrition in India: causes, consequences and possible solutions.* ISBN 978-81-322-1832-1836.

Reddy, PYB and Rao, AP 2000, 'Growth pattern of the Sugalis- a tribal population of Andhra Pradesh, India'. *Annals of Human Biology*, Vol. 27, pp. 67-81.

Rengma, MS, Bose, K and Mondal, N 2016. 'Socio-economic and demographic correlates of stunting among adolescents of Assam, North-east India'. *Anthropological Review,* Vol. 79, pp. 409-425.

Rengma, MS, Sen, J and Mondal, N 2015, 'Socio-economic, demographic and lifestyle determinants of overweight and obesity among adults of Northeast India'. *Ethiopian Journal of Health Science*, Vol. 25, pp. 199-208.

Rogol, AD 2010, 'Sex steroids, growth hormone, leptin, and the pubertal growth spurt'. *Endocrine Development*, Vol. 17, pp. 77-85.

Rogol, AD, Roemmich, JN and Clark, PA 2002, 'Growth at puberty'. *Journal of Adolescent Health*, Vol. 31, pp. S192-S200.

Rolland-Cachera, MF 1993, 'Body composition during adolescence: Method, limitation, and determinants'. *Hormone Research*, Vol. 39, pp. 25-40.

Rousham, EK, Roschnik, N, Baylon, MA, Bobrow, EA, Burkhanova, M, Campion, MG, Adle-Chua, T, Degefie, T, Hilari, C, Kalengamaliro, H, Kassa, T, Maiga, F, Mahumane, BJ, Mukaka, M, Ouattara, F, Parawan, AR, Sacko, M, Patterson, DW, Sobgo, G, Khandaker, IU and Hall, A 2011, 'A comparison of the national center for health statistics and new world health organization growth references for school-age children and adolescents with the use of data from 11 low-income countries'. *American Journal of Clinical Nutrition,* Vol. 94, pp. 571-577.

Saha, KK, Frongillo, EA, Alam, DS, Arifeen, SE, Persson, LA and Rasmussen, KM 2009, 'Use of the new world health organization child growth standards to describe the longitudinal growth of breastfed rural Bangladeshi infants and young children'. *Food and Nutrition Bulletin.*, Vol. 30, pp. 137-144.

Sawaya, AL, Martins, PA, Grillo, LP and Florencio, TT 2004, 'Long-term effects of early malnutrition on body weight regulation'. *Nutrition Review*, Vol. 62, pp. S127-S133.

Schwarz, NG, Grobusch, MP, Decker, ML, Goesch, J, Poetschke, M, Oyakhirome, S, Kombila, D, Fortin, J, Lell, B, Issifou, S, Kremsner, PG and Klipstein-Grobusch, K 2008, 'WHO 2006 child growth standards: implications for the prevalence of stunting and underweight-for-age in a birth cohort of Gabonese children in comparison to the centers for disease control and prevention 2000 growth charts and the national center for health statistics 1978 growth references'. *Public Health Nutrition*, Vol. 11, pp. 714-719.

Semproli, S and Gualdi-Russo, E 2007, 'Childhood malnutrition and growth in a rural area of Western Kenya'. *American Journal of Physical Anthropology*, Vol. 132, pp. 463-469.

Sen, J, and Mondal, N 2013, 'Fat mass and fat-free mass as indicators of body composition among Bengalee Muslim children'. *Annals of Human Biology*, Vol. 40, pp. 286-293.

Sen, J, and Mondal, N 2012, 'Socio-economic and demographic factors affecting the composite index of anthropometric failure (CIAF)'. *Annals of Human Biology*, Vol. 39, pp. 129-136.

Sharma, J and Mondal, N 2014, 'Prevalence of double nutrition burden among adolescent girls of Assam, Northeast India'. *Journal of Nepal Paediatric Society*, Vol. 34, pp. 132-137.

Sharma, J and Mondal, N 2018, Physical growth and body composition assessment among rural adolescent girls (10-16 years) of Karbi Anglong, Assam, Northeast India'. *Journal of Life Science*, Vol. 10, pp. 1-13.

Silveira, CR, Beghetto, MG, Carvalho, PR and Mello, ED 2011, 'Comparison of NCHS, CDC and WHO growth charts in the nutritional assessment of hospitalized children up to five years old'. *Nutricion Hospitalaria*, Vol. 26, pp. 465-471.

Singh, J and Mondal, N 2014, 'Use of upper-arm anthropometry as a measure of body composition and nutritional assessment in children and adolescents (6-20 years) of Assam, Northeast India'. *Ethiopian Journal of Health Science*, Vol. 24, pp. 243-252.

Singh, J, and Mondal, N 2013, 'Assessment of nutritional status case study of children in Assam, Northeast India'. *Journal of Nepal Paediatric Society*, Vol. 33, pp. 1-7.

Subramanian, SV and Smith, GD 2006, 'Patterns, distribution, and determinants of under- and overnutrition: a population-based study of women in India'. *American Journal of Clinical Nutrition,*, Vol. 84, pp. 633-640.

Svedberg, P 2011, 'How many people are malnourished?' *Annual Review of Nutrition,* Vol. 31, pp. 263-283.

Thibault, R, Genton, L and Pichard, C 2012, 'Body composition: Why, when and for who?' *Clinical Nutrition,* Vol. 31, pp. 435-447.

Thibault, R and Pichard, C 2012, 'The evaluation of the body composition: A useful practice'. *Annals of Nutrition and Metabolism*, Vol. 60, pp. 6-16.

Tigga, Pl, Sen, J and Mondal, N 2015, 'Association of some socio-economic and socio-demographic variables with wasting among preschool children of North Bengal, India'. *Ethiopian Journal of Health Science*, Vol. 25, pp. 63-72.

Touitou, Y, Portaluppi, F, Smolensky, MH and Rensing, L 2004, 'Ethical principles and standards for the conduct of human and animal biological rhythm research'. *Chronobiology International*, Vol. 21, pp. 161-170.

Ulijaszek, SJ and Kerr, DA 1999, 'Anthropometric measurement error and the assessment of nutritional status'. *British Journal of Nutrition*, Vol. 82, pp. 165-177.

Varadharajan, KS, Thomas, T and Kirpa, AV 2013, 'Poverty and state of nutrition in India'. *Asia Pacific Journal of Clinical Nutrition*, Vol. 22, pp. 326-339.

Wells, JCK 2010, *The evolutionary Biology of human body fatness: thrift and control*. Cambridge: Cambridge University Press.

WHO Multicentre Growth Reference Study Group 2006, *WHO child growth standards: length/height-for-age, weight-for-age, weight-for-length weight-for-height and body mass index-for-age: Methods and development.* Geneva: World Health Organization.

World Health Organisation 2007, *WHO Child growth standards: Head circumference-for-age, arm circumference-for-age, triceps skinfold-for-age, and subscapular skinfold-for-age: methods and development.* Geneva: World Health Organization.

World Health Organisation 1995, *Physical status the use and interpretation of anthropometry report of a WHO expert committee*. WHO, Geneva.

World Health Organization 1983, *Measuring change in nutritional status: guidelines for assessing the nutritional impact of supplementary feeding programmes for vulnerable groups.* Geneva: World Health Organization. http://www.who.int/iris/handle/10665/38768.

Xu, Y and Hang, L 2017, 'Height inequalities and their change trends in China during 1985-2010: Results from 6 cross-sectional surveys on children and adolescents aged 7-18 years'. *BioMed Central Public Health,* Vol. 17, pp. 473.

Zong, X, Li, H, Zhang, YQ and Wu, H 2017, 'Narrowing urban-suburban rural disparities in physical growth among children in China: Findings from the 5th National Survey in 2015'. *Annals of Human Biology*, Vol. 44, pp. 636-641.

Chapter - 6

Nutritional Status among the Adult Dibongiya Deori Women of Lakhimpur District, Assam

— *Juri Borah and Sarthak Sengupta*

Introduction

Health is the most fundamental facet of human life. It represents not only the bio-social development of an individual but also the society as a whole. Hence, improvement in the health status of the population has remained one of the important objectives of a country's planning towards its overall development. Nutrition is considered as one of the most significant factors that affect human health. Consequently, for proper maintenance of health throughout life adequate nutritional level is of prime importance. It has been suggested that body fat composition varies considerably between ethnic groups (Norgan, 1990; Gallagher et al., 2000). Ethnic disparity in health status is considered as the outcome of various biological, socio-cultural and economic factors. However, it is location specific also. Since various factors are responsible for the differential health status, the specific causes behind it are difficult to understand, unless studied at the micro level or household level.

Anthropometric measurements have assumed a major role in the assessment of the nutritional status of a community (Ghosh et al., 2001). Quetelet or body mass index (BMI) is the most established anthropometric indicator used for assessment of nutritional status in adults (James et al., 1988; Ferro-Luzzi et al., 1992; Shetty and James, 1994; Naidu, 1994; Bailey and Ferro-Luzzi, 1995; Lee and Nieman, 2003). Mid-upper arm circumference (MUAC) is another anthropometric measure which is particularly effective in the determination of malnutrition among adults in developing countries (James et al., 1994). The waist-to-height ratio (WHtR) has been proposed as a good measurement to identify the abdominal adiposity, particularly in individuals who would not be classified as overweight or obese by BMI (Ashwell, 2005). For the evaluation of body condition waist-to-hip ratio (WHR) is also an important indicator which represents both the subcutaneous and intra-abdominal adipose tissues (WHO, 2008). Conicity index (CI) and waist circumference (WC) also measure the body shape and are correlated with body composition. Earlier studies indicate that body mass index (BMI), waist circumference (WC) and waist-to-hip ratio (WHR) could be used independently to identify overweight and obesity (Gill et al., 2003). While, BMI is a measure of overall adiposity, waist circumference (WC), waist-to-hip ratio (WHR) and conicity index (CI) are reliable proxy measures of abdominal fat (Bose and Mascie-Taylor, 1998; Kopelman, 2000). Each of these measures has its strengths and weaknesses, and there is no consensus on which measure is the best (Freedman and Perry, 2000; Burkhauser and Cawley, 2008).

Women's status is an important factor because family's lifestyle is controlled more by their resources than by the men and Kishor and Neitzel, 1996; Haddad et al., 1997; Smith, 1998). Hence, studies on the health status of women become a focal matter of concern.

Northeast India is a multiethnic region with vast socio-economic inequality among and between these different cultural populations. Here, considerable variation between rural and urban areas with respect to health facilities, education, and infrastructural facilities are also mentioned by various scholars from time to time.

The major objectives of the present study are to examine the magnitudes of rural-urban disparity in regards to nutritional status among the adult Dibongiya Deori women of Lakhimpur district, Assam. Some socio-demographic variables are also considered that are known to affect nutritional status.

Materials and Methods

The study area and the population

The information used in this study was gathered during 2015-2016. Anthropometric measurements along with socio-demographic information were taken from Dibongiya Deori women residing in both rural (n = 258) and urban (n = 96) areas of Lakhimpur district, Assam. Rural areas considered for the present study were namely, Borchapori Deori Gaon, Kinapather Deori Gaon and Bhogpur Deori Gaon, and the different urban settlement having Dibongiya Deori concentration are Bihpuria, Bandardewa, and Narayanpur. All the respondents are ever married adult women aged between 18 years and 59 years. Women who volunteered and gave written consent were only taken into consideration.

The Dibongiya Deori is one of the sub-divisions or territorial groups (*Khels*) of Deori community. Deori, a Tibeto-Mongoloid population group, is one of the plains scheduled tribes of Assam. The concentration of Deori population can be traced from various districts of upper Assam namely, Lakhimpur, Dhemaji, Dibrugarh, Golaghat, Jorhat, Majuli, Sibsagar, eastern parts of Sonitpur and Tinsukia district and also in Lohit and Changlang districts of Arunachal Pradesh. The numeric strength of Deori population is not so much strong. In Assam, as per 2011 census report, it was recorded as 43,750 in number.

Anthropometric measurements

A portable weighing machine, anthropometer and steel tape were used for the measurements. All the measurements were taken following the standard techniques. Height and weight measurements were taken on each subject using standard protocols given by Weiner and Lourie (1981). Waist circumference (WC) was measured midway between the iliac crest and the lowermost margin of the ribs (WHO, 2000 cut off value). Hip circumference was measured at the maximum circumference of the buttocks, with the subject standing and his feet placed together. Mid-upper arm circumference (MUAC) were taken at the middle (midway between acromion and elbow) part of the left upper arm, while the arms were hanging at the sides of the body (Ferro-Luzzi and James, 1996 cut off value).

Besides the above measurements, following indices and ratios were computed:

1. Body mass index or BMI= weight(kg)/height(m)^2 (WHO, 2004 cut off value)
2. Waist-to-Hip ratio (WHR)= waist circumference(cm)/hip circumference(cm) (Webb, 2002)
3. Waist-to-Height ratio (WHtR)=waist circumference(cm)/ height(cm) (Hsieh and Muto, 2004)
4. Conicity index (CI)= waist circumference(m)/ 0.109x$\sqrt{}$ {weight(kg)/height(m)}

Statistical analyses

Descriptive statistics including means, standard deviation, standard error, and percentage were calculated. To compare the mean values of continuous variables between rural and urban areas, Student's t-test was used and Chi-square test for categorical variables. A significance value of 0.5 percent was considered for this study. All the analyses were done by using the computer software namely, Chi-Square Test of Independence and Microsoft Office Excel 2007. Technical errors of measurements were found to be within acceptable limits (Ulijaszek and Kerr, 1999) and therefore, were not incorporated in the analyses.

Findings and Discussion

The Dibongiya Deori women are classified in the age groups of ≤20-29 years, 30-39 years, 40-49 years and 50-59 years. In the rural area, more numbers of women are found in the age group of ≤20-29 years (31.01%), followed by the age group of 40-49 years (29.07%), 30-39 years (26.36%) and 50-59 years (13.57%). On the other hand, the distribution of Dibongiya Deori women in the urban area is found as comparatively higher in the age group of 30-39 years (31.25%), followed by the age group of 40-49 years (25.00%), 50-59 years (22.92%) and ≤20-29 years (20.83%). Regarding educational status, the proportions of the illiterate women are found as considerably higher in the rural area (20.16%) in comparison to the women in the urban area (3.13%). Moreover, noticeably higher proportions of the women from an urban area (69.79%) than their rural counterparts (35.66%) have beyond high school level educational profile. 10.85 percent of women from a rural area and 4.17 percent of the women from the urban area have attained education up to primary standard. However, quite a good numbers of women from rural (33.33%) and urban (22.92%) area have educational attainment up to high school level. Most of the Dibongiya Deori women from rural (93.02%) and urban (78.13%) area are not employed outside the home and remain as homemakers. However, the proportions of the working women are found as considerably higher in the urban area (21.88%) in comparison to their rural counterparts (6.98%).

Table 6.1 displays the means of anthropometric measurements and indices among the ever-married Dibongiya Deori women. The mean of all anthropometric measurements and indices of the urban area considered are found as significantly higher than their rural counterparts except the WHR. In regards to WHR, almost equal consistency is noticed between the women from both rural and urban setup. In the study of Temsutola (2010) among the Ao adult women of Nagaland also found comparatively greater mean anthropometric measurements and indices such as height, weight, MUAC, WC, Hip circumference, CI and BMI in the urban area than the rural group of women. She (2010) recorded a little higher mean WHR in the rural area than their urban counterpart among the Ao adult women of Nagaland.

Table 6.1: Anthropometric measurements and indices among the ever-married women

Measurements and Indices	*Mean ± S.E.*		*t-value*
	Rural (n= 258)	*Urban (n= 96)*	
Height (cm)	152.91 ± 0.32	154.04 ± 0.42	2.14
Weight (kg)	50.22 ± 0.50	58.14 ± 1.00	7.08
Waist Circumference (cm)	76.91 ± 0.50	85.58 ± 1.01	7.69
Hip Circumference (cm)	90.09 ± 0.55	98.91 ± 0.87	7.77
Mid Upper Arm Circumference (cm)	26.58 ± 0.42	29.29 ± 0.77	3.09
Body Mass Index (BMI)	21.45 ± 0.19	24.50 ± 0.41	6.75
Waist Hip Ratio (WHR)	0.85 ± 0.002	0.86 ± 0.01	0.98
Waist Height Ratio (WHtR)	0.50 ± 0.003	0.56 ± 0.01	5.75
Conicity Index (CI)	1.23 ± 0.004	1.28 ± 0.01	4.64

In regards to mean height of the present Dibongiya Deori women from rural area (152.91 cm), conformity is observed with that of ethnic groups living in the rural area such as Dibongiya Deori (153.28 cm; Gogoi, 2006), Mishing (153.07 cm; Dutta and Sengupta, 2008) and Boro Kachari (152.07 cm; Sengupta and Hussain, 2014) of Assam and also Ao of Nagaland (152.72 cm; Temsutola, 2010). Comparatively lower mean height than the present Dibongiya Deori women from rural area is observed among the Bengali (150.03 cm; Sikdar and Ahmed Das, 2006, calculated by JB), Kalita (151.53 cm; Dutta and Sengupta, 2008) and Bishnupriya Manipuris (149.37 cm; Das and Roy, 2010, calculated by JB) of Assam and the Tangkhul Naga of Manipur (151.30 cm; Mungreiphy and Kapoor, 2010). The mean height of Tengaponiya (154.41 cm) and Borgonya Deori (156.21 cm) studied by Gogoi (2006) and Sonowal Kachari (154.44 cm; Dutta and Sengupta, 2008) of Assam and Nyishi of Arunachal Pradesh (154.73 cm; Bharali et al., 2017) are found to be higher than the Dibongiya Deori women from rural area. The mean height of Dibongiya Deori women from urban area is found as 154.04 cm which is greater than the mean value found among the Karbi residing in the semi-urban area of Assam (152.52 cm; Goswami and Bhattacharyya, 2015) and other urban residents like Ao of Nagaland (153.30 cm; Temsutola, 2010) and Meitei of Manipur (153.2 cm; Singh, 2013).

The mean weight of the Dibongiya Deori women from a rural area is recorded as 50.22 kg which shows similarity with that of Tengaponiya Deori (50.22 kg; Gogoi, 2006), Kalita 50.56 kg; Dutta and Sengupta, 2008) and Boro Kachari (50.66 kg; Sengupta and Hussain, 2014) of Assam. The mean weight of Bengali (52.11 kg; Sikdar and Ahmed Das, 2006, calculated by JB) and Khamyang (55.15 kg; Dutta and Sengupta, 2011) of Assam and Nyishi of Arunachal Pradesh (51.00 kg; Bharali et al., 2017) has higher mean weight than the present population from a rural area. On the other hand, the population groups in Assam like Borgonya (48.90 kg) and Dibongiya Deori (47.50 kg) studied by Gogoi (2006), Sonowal Kachari (49.19 kg) and Mishing (48.65 kg) studied by Dutta and Sengupta (2008) and Bishnupriya Manipuris (47.15 kg; Das and Roy, 2010, calculated by JB); Ao of Nagaland (47.74 kg; Temsutola, 2010) and Tangkhul Naga of Manipur (48.68 kg; Mungreiphy and Kapoor, 2010) have lower mean weight than the Dibongiya Deori women of rural area. The mean weight of Dibongiya Deori women of urban area (58.14 kg) is found as considerably higher than the population groups from semi-urban area like Karbi in Assam (50.53 kg; Goswami and Bhattacharyya, 2015) and urban areas that is Ao of Nagaland (49.10 kg; Temsutola, 2010) and Meitei of Manipur (54.7 kg; Singh, 2013).

Mean value of MUAC among the Dibongiya Deori women is recorded as 26.58 cm and 29.29 cm in the rural and urban area respectively. The mean MUAC in the rural area is found as comparatively higher than so far reported mean MUAC among the Kalita (25.81cm) and Mishing (25.40 cm) studied by Dutta and Sengupta (2008), Bishnupriya Manipuris (22.45 cm; Das and Roy, 2010, calculated by JB) and Boro Kachari (25.52 cm; Sengupta and Hussain, 2014) in Assam and Ao of both rural (22.26 cm) and urban (22.78 cm) areas of Nagaland (Temsutola, 2010). The Sonowal Kachari women (27.11cm) in Assam shows moderately higher mean MUAC than the present rural Dibongiya Deori women in Assam.

The mean BMI of Dibongiya Deori from a rural area is 21.45 and 24.50 in the urban area. In regards to the mean BMI of rural Dibongiya Deori women, similarity is observed with that of Mishing (21.30; Basu, 2003 and 21.94; Baruah, 2007), Tengaponiya Deori (20.95; Gogoi, 2006), Sonowal Kachari (21.24; Baruah, 2007), Kalita (21.95; Dutta and Sengupta, 2008) and Bishnupriya Manipuris (21.11; Das and Roy, 2010, calculated by JB) in Assam, Nyishi tribal women of Arunachal Pradesh (21.30; Bharali et al., 2017) and Tangkhul Naga women of Manipur (21.23; Mungreiphy and Kapoor, 2010). In Assam, the Bodo (20.67; Basu and Gajbhieye, 1999),

Borgonya (20.05) and Dibongiya Deori (20.20) studied by Gogoi (2006), Ahom (20.63; Baruah, 2007), Sonowal Kachari (20.68) and Mishing (20.85) studied by Dutta and Sengupta (2008) and Ao of rural areas of Nagaland (20.46; Temsutola, 2010) show relatively lower mean BMI than the present Dibongiya Deori women. The Dimasa (22.02; Basu, 2006) and Khamyang (23.89; Dutta and Sengupta, 2011) in Assam show markedly higher mean BMI in comparison to the Dibongiya Deori women from rural areas. However, the respondents from a semi-urban area that is Karbi in Assam (21.53; Goswami and Bhattacharyya, 2015) and urban areas namely Ao of Nagaland (20.89; Temsutola, 2010) and Meitei of Manipur (23.3; Singh, 2013) have lower mean BMI than the Dibongiya Deori women from urban areas.

The mean WHR of the Dibongiya Deori women from a rural and urban area is recorded as 0.85 and 0.86 respectively. Both these mean values are found as slightly higher than the Ao women of rural (0.83) and urban (0.82) areas of Nagaland (Temsutola, 2010). The Kalita (0.87), Sonowal Kachari (0.89) and Mishing (0.89) studied by Dutta and Sengupta (2008) found comparatively higher WHR in comparison to the present Dibongiya Deori women.

The mean WC of the present sample is recorded as 76.91 cm in a rural area and 85.58 cm in an urban area. In Assam, the WC of rural areas is found as considerably lower than the Bengali (83.26 cm; Sikdar and Ahmed Das, 2006) and Kalita (82.17 cm), Sonowal Kachari (80.04 cm) and Mishing (79.36 cm) studied by Dutta and Sengupta (2008). Conversely, these values are found as considerably higher than Ao women in rural (65.29 cm) and urban (66.83 cm) area of Nagaland (Temsutola, 2010).

Regarding Hip circumference, the rural Dibongiya Deori (90.09 cm) shows conformity with the Sonowal Kachari (90.06 cm) and Mishing (89.31 cm) in Assam (Dutta and Sengupta, 2008). The Kalita women in Assam (93.70 cm; Dutta and Sengupta, 2008) has considerably higher mean hip circumference than the present Dibongiya Deori women from the rural area. The mean hip circumference of Dibongiya Deori women from both rural (90.09 cm) and urban areas (98.91cm) is markedly higher than the value obtained from Ao women in rural (78.71cm) and urban (81.00 cm) area of Nagaland (Temsutola, 2010).

In regards to the mean CI of Dibongiya Deori from rural (1.23) and urban (1.28) areas, the similarity is observed with that of Kalita (1.31), Sonowal Kachari (1.30) and Mishing (1.30) of Assam (Dutta and Sengupta, 2008). The present mean CI irrespective of their residence is found to be noticeably higher than the Ao women of rural (1.07) and urban (1.07) areas in Nagaland (Temsutola, 2010).

Table 6.2 shows the nutritional status according to BMI among the ever-married Dibongiya Deori women. More numbers of overweight (32.29%) and obese (7.29%) mothers are found from urban area whereas among the rural area the frequency of undernutrition (15.12%) is relatively common. The frequency of the women that belongs to the normal category is considerably lower in the urban group of women (56.25%) than their rural counterparts (70.54%). The test of significance shows statistically significant differences between the rural and urban groups of women. In various rural areas of Assam, considerably higher proportions of underweight women are also recorded among the Karbi (37.20%; Kulkarni and Sharma, 2001), Ahom (23.78%; Baruah, 2007), Sonowal Kachari (18.37%; Baruah, 2007 and 20.19%; Dutta and Sengupta, 2008), Mishing (23.76%; Dutta and Sengupta, 2008) and Meitei (33.3%; Islam Barbhuiya and Das, 2013) of Assam. Among the Tangkhul Naga women of Nagaland also Mungreiphy and Kapoor (2010) reported the presence of more underweighted (16.00%) than the overweight (9.2%) and obese (0.6%) category. Similar finding that is more prevalence of undernourished women in the rural area in contrast to overweight and obesity in higher proportions in an urban area was also found by NFHS IV (2015-2016) report for Assam as a whole.

Table 6.2: Nutritional status according to BMI among the ever-married women

Nutritional status (BMI)	*Rural (n= 258)*		*Urban (n= 96)*		χ^2
	No.	*%*	*No.*	*%*	
Underweight (<18.50)	39	15.12	4	4.17	33.90 p = 0.0001*
Normal (18.51-24.99)	182	70.54	54	56.25	
Overweight (25.00-29.99)	35	13.57	31	32.29	
Obese (≥30.00)	2	0.78	7	7.29	
Total	258	100.00	96	100.00	

Classification of WC, WHR, WHtR, CI, and MUAC among the ever-married Dibongiya Deori women is shown in Table 6.3. Regarding WC and WHtR, significantly higher proportions of the women from an urban area (WC=64.58%, WHtR=83.33%) are found to be obese than the women from rural areas (WC=30.62%, WHtR=48.84%). On the contrary, the frequencies of obese women by WHR are observed as higher in the rural areas (92.25%) in comparison to the urban group of women (88.54%), although the difference is not found as statistically significant as per chi-square test of significance. Women categorized as undernourished as per MUAC is marginally higher in the rural area (3.10%) than the urban area (2.08%). Likewise, a slightly greater level of CI is found in women from a rural area (98.06%) in comparison to the women from an urban area (97.92%). Hence, no statistically significant differences are observed in regards to MUAC and CI between the two samples considered.

Table 6.3: Classification of WC, WHR, WHtR, CI and MUAC among the ever-married women

Variables	*Category*	*Rural (n= 258)*		*Urban (n= 96)*		χ^2
		No.	*%*	*No.*	*%*	
WC	Normal (< 80)	179	69.38	34	35.42	33.68
	High/Obese (≥ 80)	79	30.62	62	64.58	p = 0.0001*
WHR	Normal (≤ 0.80)	20	7.75	11	11.46	1.20
	High/Obese (> 0.80)	238	92.25	85	88.54	p = 0.27
WHtR	Normal (< 0.50)	132	51.16	16	16.67	34.22
	High/Obese (≥ 0.50)	126	48.84	80	83.33	p = 0.0001*
CI	< 1.10	5	1.94	2	2.08	0.008
	> 1.10	253	98.06	94	97.92	p = 0.93
MUAC	Normal (> 22)	250	96.90	94	97.92	0.26
	Under nutrition (< 22)	8	3.10	2	2.08	p = 0.61

The relationship between BMI and background characteristics is shown in Table 6.4. In the rural area, noticeably higher proportions of the women are found as undernourished in the age group of ≤ 20-29 years (48.75%). In the urban area also, the prevalence of underweight women is comparatively higher in this age group (10.00%). Not a single woman is found as obese in the age group of ≤ 20-29 years in both the setup. In the rural area, the frequency of woman belongs to the overweight category is found as markedly higher in the age group of 50-59 years (94.29%). In the case of urban area, noticeable proportions of the women are found as overweight in the age group of ≤ 20-29 years (20.00%) also. However, comparatively higher frequencies of overweight women in the urban area are observed in the age group of 30-39 years (43.33%), followed by the age group of 50-59 years (36.36%). In the urban area, the prevalence of obese women is recorded as comparatively higher in the age group of 40-49 years (16.67%), while it is recorded only in the age group of 50-59 years (5.71%) in case of rural area. In regard to age groups, the test of significance shows statistically significant differences between the rural and urban groups of women. The prevalence of underweight women in higher proportions among the youngest age group (20–29 years) was also reported by Mungreiphy and Kapoor (2010) among the Tangkhul Naga women of Nagaland. Among them, they observed the prevalence of overweight women in higher

Table 6.4: The relationship between BMI and background characteristics among the ever-married women

Background characters	*Nutritional status (BMI)*										χ^2
	Rural					*Urban*					
	Under weight	*Normal*	*Over weight*	*Obese*	*Total*	*Under weight*	*Normal*	*Over weight*	*Obese*	*Total*	
Present age (in years)											
≤20-29	39 (48.75)	41 (51.25)	0	0	80	2 (10.00)	14 (70.00)	4 (20.00)	0	20	19.68,p= 0.0001*
30-39	0	68 (100.00)	0	0	68	1 (3.33)	14 (46.67)	13 (43.33)	2 (6.67)	30	32.24,p= 0.0001*
40-49	0	73 (97.33)	2 (2.67)	0	75	0	14 (58.33)	6(25.00)	4 (16.67)	24	23.84,p= 0.0001*
50-59	0	0	33 (94.29)	2 (5.71)	35	1(4.55)	12 (54.55)	8(36.36)	1 (4.55)	22	19.04,p= 0.0001*
Educational level											
Illiterate	6 (11.54)	37 (71.15)	9 (17.31)	0	52	0	2 (66.67)	1 (33.33)	0	3	0.43,p= 0.93
Primary	2 (7.14)	22 (78.57)	2 (7.14)	2 (7.14)	28	0	2 (50.00)	2 (50.00)	0	4	5.60,p= 0.13
Up to high school	19 (22.09)	58 (67.44)	9 (10.47)	0	86	1 (4.55)	15 (68.18)	4 (18.18)	2 (9.09)	22	10.01,p= 0.02*
Beyond high school	12 (13.04)	65 (70.65)	15 (16.30)	0	92	3 (4.48)	35 (52.24)	24 (35.82)	5 (7.46)	67	15.10,p= 0.002*
Occupational level											
Working	1 (5.56)	14 (77.78)	3 (16.67)	0	18	1 (3.23)	11 (35.48)	6 (19.35)	3 (9.68)	21	2.77,p= 0.43
Non working	38 (15.83)	168 (70.00)	32 (13.33)	2 (0.83)	240	3 (4.62)	43 (66.15)	25(38.4 6)	4 (6.15)	75	26.22,p= 0.0001*

proportions among the three middle-age groups, i.e. 30–39 years, 40–49 years and 50–59 years. NFHS IV (2015-2016) report for Assam as a whole also documented the common occurrence of undernutrition in younger age groups and overweight and obesity among the older adults. Among the Karbi women residing the semi-urban area in Assam, Goswami and Bhattacharyya (2015) observed the higher prevalence of undernutrition in the young age group (19-29 years) and overweight in the higher age group (40-49 years). Bordoloi and Kapoor (2013) among the Kalita women of Assam also observed a higher proportion of undernourished women as per BMI in the age group of 20-29 years and overweight women in the age group of 40-49 years.

In both rural and urban areas, the proportions of underweight women are found to be higher among those whose has attained education up to high school level (rural = 22.09% and urban = 13.04%), followed by the women with beyond high school level educational profile (rural = 4.55% and urban = 4.48%). Comparatively higher frequency of overweight women is observed in the illiterate section (17.31%) in the rural area. However, modest proportions of the women from urban area belonging to the primary level educational category (50.00%) are found as overweight. In the illiterate category irrespective of their residence, the frequency of obese women is recorded as nil. In the rural area, obese women are found only among those who have attained education up to primary level (7.14%). On the contrary, comparatively number of women having education up to high school level (9.09%) is found as obese, followed by those who have beyond high school standard educational attainment (7.46%). As per educational level, the chi-square test of significance records statistically significant differences between the two samples considered except the illiterates and the women having education up to primary level. Among the Meitei of urban areas of Manipur (Singh, 2013) also found a higher prevalence of overweight women having primary education. However, the prevalence of obesity was found as higher among the Meitei women having education up to higher secondary level and beyond that.

Regarding the occupational level, the proportions of underweight women are comparatively higher in the non-working category in both rural (15.83%) and urban (4.62%) areas. Although in the working category also the prevalence of overweight women is quite high irrespective of their place of residence, but the occurrence of it is considerably higher in the non working category (38.46%) in the urban area. In the case of a rural area, the prevalence of obese women is found only in the non-working category (0.83%). Conversely, in the urban area, the frequency of obese women is found as marginally higher in the working category (9.68%). In this regard, statistically significant differences between rural and urban areas are observed in the non-working category only. Among the Tangkhul Naga women of Nagaland (Mungreiphy and Kapoor, 2010) also found the greatest numbers of overweight and obese subjects among the homemakers.

Table 6.5 shows the relationship between WC and background characteristics among the ever-married Dibongiya Deori women. The proportion of obese women by waist circumference is relatively higher in the age group of 50-59 years in both rural (42.86%) and urban (77.23%) areas. In this regard, the subsequent declining trend is noticed as the ages of the respondents come down in the rural area. In the urban area also the proportion of obese women is found as lower in the age group of $\leq$ 20-29 years. However, statistically significant differences between the two samples are recorded as per the chi-square test of significance except for the age group of $\leq$ 20-29 years.

The proportions of overweight women are noticeably higher among the illiterates (44.23%) in the rural area. In the urban area, considerably higher numbers of women having education up to primary standard are found as obese (75.00%). Interestingly, the proportions of obese women in the urban area are found as low among the illiterates (33.33%). The test of significance shows significant differences between the two studied samples with regards to their educational attainment except the women having education up to primary level.

In both the rural (44.44%) and urban (80.95%) area, considerably higher proportions of obese women are found in the working category. In this connection, statistically significant differences between rural and urban women are also observed according to the test of significance.

Relationship of WHR with background characteristics is shown in Table 6.6. A comparatively higher number of obese women according to WHR is found in the age group of 50-59 years in both rural (97.14%) and urban (100.00%) areas. In this regard, the chi-square test of significance records statistically significant differences in the age group of $\leq$ 20-29 years only between the two samples considered.

All the illiterate, as well as the women having education up to primary standard, are found as obese in the urban area. In the rural area also, comparatively higher proportions of obese women are found in the illiterate category. The variation observed in between the rural and urban samples, statistically significant only among the women with beyond high school level educational profile.

Table 6.5 : The relationship between WC and background characteristics among the ever-married women

Background characters	*Nutritional status (WC)*						χ^2
	Rural			*Urban*			
	Normal	*High/Obese*	*Total*	*Normal*	*High/Obese*	*Total*	
Present age (in years)							
≤20-29	66 (82.5)	14 (17.5)	80	13 (65.00)	7 (35.00)	20	2.95 p= 0.09
30-39	50 (73.53)	18 (26.47)	68	8 (26.67)	22 (73.33)	30	18.92, p= 0.0001*
40-49	43 (57.33)	32 (42.67)	75	8 (33.33)	16 (66.67)	24	4.19, p= 0.04*
50-59	20 (57.14)	15 (42.86)	35	5 (22.73)	17 (77.23)	22	6.50, p= 0.01*
Educational level							
Illiterate	29 (55.77)	23 (44.23)	52	2 (66.67)	1 (33.33)	3	11.58, p= 0.001*
Primary	20 (71.43)	8 (28.57)	28	1 (25.00)	3 (75.00)	4	3.34, p= 0.07
Up to high school	65 (75.58)	21 (24.42)	86	7 (31.82)	15 (68.18)	22	15.10, p= 0.0001*
Beyond high school	65 (70.65)	27 (29.35)	92	24 (35.82)	43 (64.18)	67	19.09, p= 0.0001*
Occupational level							
Working	10 (55.56)	8 (44.44)	18	4 (19.05)	17 (80.95)	21	5.61, p= 0.02*
Non working	169 (70.42)	71 (29.58)	240	30 (40.00)	45 (60.00)	75	22.73, p= 0.001*

Figure within parentheses indicates the percentage.

Table 6.6: The relationship between WHR and background characteristics among the ever-married women

Background characters	*Nutritional status (WHR)*						χ^2
	Rural			*Urban*			
	Normal	*High/Obese*	*Total*	*Normal*	*High/Obese*	*Total*	
Present age (in years)							
≤20-29	8 (10.00)	72 (90.00)	80	7 (35.00)	13 (65.00)	20	7.84, p= 0.005*
30-39	6 (8.82)	62 (91.18)	68	2 (6.67)	28 (93.33)	30	0.13, p=0.72
40-49	5 (6.67)	70 (93.33)	75	2 (8.33)	22 (91.67)	24	0.08, p=0.78
50-59	1 (2.86)	34 (97.14)	35	0	22 (100.00)	22	0.25, p=0.61
Educational level							
Illiterate	2 (3.85)	50 (96.15)	52	0	3 (100.00)	3	0.01, p=0.92
Primary	3 (10.71)	25 (89.29)	28	0	4 (100.00)	4	0.10, p=0.75
Up to high school	10 (11.63)	76 (88.37)	86	1 (4.55)	21 (95.45)	22	0.96, p=0.33
Beyond high school	5 (5.43)	87 (94.57)	92	10 (14.93)	57 (85.07)	67	4.09, p=0.04*
Occupational level							
Working	0	18 (100.00)	18	0	21 (100.00)	21	0.0001, p=1.000
Non working	20 (8.33)	220 (91.67)	240	11 (14.67)	64 (85.33)	75	2.58, p=0.11

Figure within parentheses indicates the percentage

All the working women in both rural (100.00%) and urban (100.00%) area are found as obese as per WHR under the present study. However, no statistically significant differences are observed between the two samples studied.

Table 6.7 shows the relationship between WHtR and background characteristics among the ever-married Dibongiya Deori women. Relatively higher proportions of obese women are found in the age group of 50-59 years in both the rural (65.71%) and urban (95.45%) areas. In this connection, a statistically significant difference between a rural and urban group of women are observed in the age groups of 30-39 years and 50-59 years.

In the rural area, considerably higher proportions of obese women are found among the illiterates (67.31%), followed by the women having education up to primary level (57.14%). Conversely, somewhat higher proportions of women having education up to high school standard (95.45%) are found as obese in the urban area, which is followed by the women that have attained education beyond high school standard (80.60%). The variation observed between the rural and urban samples is statistically significant among the women having education up to high school level and beyond high school level.

Regarding occupational level, comparatively higher proportions of working women are found as obese in both rural (55.56%) and urban (95.24%) areas. However, the chi-square test of significance also shows statistically significant differences between the two studied samples.

Table 6.8 depicts the relationship between CI and the background characteristics among the ever married Dibongiya Deori women. All the respondents aged 40 years and above have the conicity index of >1.10 in the rural area. On the contrary, in the urban area, all the respondents less than 40 years of age are found to have the conicity index of >1.10 Among all the age groups considered in the present study, statistically, significant differences are observed in the age group of 40-49 years between the two samples.

All the illiterate women are found to have the conicity index of >1.10 in the rural area. However, in the urban area, except the women who have an educational profile beyond high school standard have the conicity index of <1.10. The test of significance fails to record any statistically significant variation in this regard between the two samples considered.

Table 6.7: The relationship between WHtR and background characteristics among the ever-married women

Background characters	*Nutritional status (WHtR)*						χ^2
	Rural			*Urban*			
	Normal	*High/Obese*	*Total*	*Normal*	*High/Obese*	*Total*	
Present age (in years)							
≤20-29	54 (67.5)	26 (32.5)	80	11 (55.00)	9 (45.00)	20	1.10,p=0.30
30-39	39 (57.35)	29 (42.65)	68	2 (6.67)	28 (93.33)	30	21.98,p=0.0001*
40-49	27 (36.00)	48 (64.00)	75	2 (8.33)	22 (91.67)	24	6.72,p=0.01*
50-59	12 (34.29)	23 (65.71)	35	1 (4.55)	21 (95.45)	22	6.79,p=0.009*
Educational level							
Illiterate	17 (32.69)	35 (67.31)	52	1 (33.33)	2 (66.67)	3	0.001,p=0.98
Primary	12 (42.86)	16 (57.14)	28	1 (25.00)	3 (75.00)	4	0.46,p=0.50
Up to high school	52 (60.47)	34 (39.53)	75	1 (4.55)	21 (95.45)	22	21.92,p=0.001*
Beyond high school	51 (55.43)	41 (44.57)	92	13 (19.40)	54 (80.60)	67	20.93,p=0.001*
Occupational level							
Working	8 (44.44)	10 (55.56)	18	1 (4.76)	20 (95.24)	21	8.60,p=0.003*
Non working	124 (51.67)	116 (48.33)	240	15 (20.00)	60 (80.00)	75	23.24,p=0.0001*

Figure within parentheses indicates the percentage

Table 6.8 : The relationship between CI and background characteristics among the ever-married women

Background characters	*Nutritional status (CI)*						χ^2
	Rural			*Urban*			
	< 1.10	*> 1.10*	*Total*	*< 1.10*	*> 1.10*	*Total*	
Present age (in years)							
≤20-29	4 (5.00)	76 (95.00)	80	0	20 (100.00)	20	0.24,p=0.62
30-39	1 (1.47)	67 (98.53)	68	0	30 (100.00)	30	0.14,p=0.71
40-49	0	75 (100.00)	75	2 (8.33)	22 (91.67)	24	4.86,p=0.03*
50-59	0	35 (100.00)	35	0	22 (100.00)	22	0.0001,p=1.000
Educational level							
Illiterate	0	52 (100.00)	52	0	3 (100.00)	3	0.0001,p=1.000
Primary	1 (3.57)	27 (96.43)	28	0	4 (100.00)	4	0.02,p=0.88
Up to high school	1 (1.16)	85 (98.84)	75	0	22 (100.00)	22	0.06,p=0.82
Beyond high school	3 (3.26)	89 (96.74)	92	2 (2.99)	65 (97.01)	67	0.01,p=0.92
Occupational level							
Working	0	18 (100.00)	18	0	21 (100.00)	21	0.0001,p=1.000
Non working	5 (2.08)	235 (97.92)	240	2 (2.67)	73 (97.33)	75	0.09,p=0.77

Figure within parentheses indicates the percentage

All the working women are found to have the conicity index of >1.10 irrespective of their place of residence. Here too, the test of significance shows insignificant differences between the two samples studied.

Relationship of MUAC with background characteristics among the ever-married Dibongiya Deori women is shown in Table 6.9. Comparatively higher proportions of undernourished women as per MUAC are found in the age group of ≤ 20-29 years in both the rural (7.5%) and urban (5.00%) area. Regarding educational attainment, fairly higher proportions of illiterate women are found as underweight in the rural area (7.69%), while in the urban area, underweight women are found among those who have educational attainment beyond high school standard (2.99%). Relatively higher frequency of underweight women by MUAC is found in the working category (4.76%) in the urban area, whereas, in the rural area, under-nourished women are observed only in the non-working category (3.33%). With regards to the background characters, the test of significance records statistically insignificant difference between the rural and urban groups of women.

Table 6.9: The relationship between MUAC and background characteristics among the ever-married women

Background characters	*Nutritional status (MUAC)*						χ^2
	Rural			*Urban*			
	Normal	*Under nutrition*	*Total*	*Normal*	*Under nutrition*	*Total*	
Present age (in years)							
≤20-29	74 (92.5)	6 (7.5)	80	19 (95.00)	1 (5.00)	20	0.15,p= 0.70
30-39	68 (100.00)	0	68	30 (100.00)	0	30	0.0001,p=1.000
40-49	75 (100.00)	0	75	23 (95.83)	1 (4.17)	24	2.40,p=0.12
50-59	33 (94.29)	2 (5.71)	35	22 (100.00)	0	22	0.53,p=0.47
Educational level							
Illiterate	48 (92.31)	4 (7.69)	52	3 (100.00)	0	3	0.03,p=0.86
Primary	28 (100.00)	0	28	4 (100.00)	0	4	0.0001,p=1.000
Up to high school	84 (97.67)	2 (2.33)	75	22 (100.00)	0	22	0.11,p=0.74
Beyond high school	90 (97.83)	2 (2.17)	92	65 (97.01)	2 (2.99)	67	0.10,p=0.75
Occupational level							
Working	18 (100.00)	0	18	20 (95.24)	1 (4.76)	21	0.42,p=0.52
Non working	232 (96.67)	8 (3.33)	240	74 (98.67)	1 (1.33)	75	0.82,p=0.36

Figure within parentheses indicates the percentage

Temsutola (2010) also reported a greater mean value of MUAC, WC, CI and WHR among the Ao women of rural and urban areas of Nagaland belong to older age groups. A higher proportion of women considered as risk group as per WC, WHR and WHtR were also found in the higher age groups among the Kalita women of Assam (Bordoloi and Kapoor, 2013).

Conclusion

The presence of overweight and obese women is quite common in the older age groups among the Dibongiya Deori women in Assam. Among the illiterate section in both the rural and urban area, a number of women considered as obese by BMI, WC, WHR, WHtR, CI, and MUAC are recorded. However, the working women irrespective of their residence are found to be relatively more sufferers with the problem of obese body composition. In urban areas, more numbers of women are found to be overweight and obese in comparison to the rural area according to their BMI. However, they are also found as more risk group in comparison to the women from a rural area as per WC and WHtR. Sedentary lifestyles with less physical activity may be one of the reasons behind this. Among the rural resident's presence of undernourished women are found to be high. Therefore proper nutrition and dietary practices, support, counseling, and health education are important to overcome this problem. More scientific works to assess the severity of these problems are of utmost importance in filling knowledge gaps and for the improvement of the health of the population which would markedly bring an effective environment in the society as a whole.

Acknowledgement

The author are grateful to all the Dibongiya Deori informants participated in this study.

References

Bailey, KV & Ferro-Luzzi, A 1995, 'Use of body mass index of adults in assessing individual and community nutritional status', *World Health Organization Bulletin*, Vol. 73, no. 5, pp. 673–680.

Baruah, P 2007, *Demographic profile of three ethnic groups of Assam: a comparative study.* Ph.D. thesis, Department of Anthropology, Dibrugarh University.

Basu, D & Gajbhieye, S 1999, *Nutritional status of Indian population: the Mishing and Bodo of Sonitpur district of Assam*. Report submitted to Anthropological Survey of India, Government of India.

Basu, D 2003, 'Nutritional anthropometric status among the Mishing adults of Sonitpur district, Assam'. *Paper presented in the Golden Jubilee Seminar on "Anthropological Perspectives: Northeast India" from 28th to 30th April 2003*, Anthropological Survey of India, Shillong.

Basu, D 2006, 'Nutritional anthropometric assessment of nutrient's intakes in the Dimasa Kachari adult population of Assam' in *Agriculture food security, nutrition, and health in N.E. India*, eds D Basu, B Francis Kulirani & B Datta Roy, Akansha Publications, New Delhi.

Bharali, N, Mondal, N & Singh, KhN 2017, 'Prevalence of undernutrition, overweight and obesity among Nyishi tribal women of Arunachal Pradesh, Northeast India', *Human Biology Review*, Vol. 6, no. 1, pp. 63-78.

Bordoloi, T & Kapoor, A 2013, 'Prevalence of cardiovascular risk factors with ageing: a study in a biologically isolated group of Northeast India', *Asian Journal of biological and Life Sciences*, Vol. 2, no. 2, pp. 114-118.

Bose, K & Mascie-Taylor, CGN 1998, 'Conicity index and waist-to-hip ratio and their relationship with total cholesterol and blood pressure in middle-aged Europeans and migrant Pakistani men', *Annals of Human Biology*, Vol. 25, pp. 11-16.

Burkhauser, RV & Cawley, J 2008, 'Beyond BMI: the value of more accurate measures of fatness and obesity in social science research', *Journal of Health Economics*, Vol. 27, pp. 519-529.

Das, BM & Roy, S 2010, 'Age changes in the anthropometric and body composition characteristics of the Bishnupriya Manipuris of Cachar district, Assam', *Advances in Bioscience and Biotechnology*, Vol. 1, pp. 122-130.

Dutta, D & Sengupta, S 2008, 'Prevalence of overweight and obesity among three ethnic groups of Assam: a preliminary appraisal', *Bulletin of the Department of Anthropology, Dibrugarh University*, Vol. 40, pp. 1-16.

Dutta, D & Sengupta, S 2011, 'Somatometric characters of the Khamyangs of Assam', *Bulletin of the Department of Anthropology, Dibrugarh University*, Vol. 39, pp. 61-68.

Ferro-Luzzi, A & James, WP 1996, 'Adult malnutrition: simple assessment techniques for use in emergencies', *British Journal of Nutrition*, Vol. 75, no. 1, pp. 3-10.

Ferro-Luzzi, A, Sett'e, S & Franklin, M 1992, A simplified approach of assessing adult chronic energy deficiency', *European Journal of Clinical Nutrition*, Vol. 46, pp. 173–186.

Freedman, DS & Perry, G 2000, 'Body composition and health status among children and adolescents', *Preventive Medicine*, Vol. 31, no. 2, pp. 34-53.

Gallagher, D, Heymsfield, SB, Heo, M, Jebb, SA, Murgatroyd, PR & Sakamoto, Y 2000, 'Healthy percentage body fat ranges an approach for developing guidelines based on body mass index', *American Journal of Clinical Nutrition*, Vol. 72, pp. 694–701.

Ghosh, R, Das, PK & Bharati, P 2001, 'Health and nutritional status of Ho preschool children of Orissa', *Journal of Human Ecology*, Vol. 12, pp. 109–113.

Gill, T, Littleborough, C & Taylor, A 2003, 'Body mass index, waist-hip ratio, and waist circumference: which measure to classify obesity?', *Soz Preventivemed*, Vol. 48, no. 191-200.

Gogoi, G 2006, *A study on population genetics among the Deori of Assam*, Ph.D. thesis, Department of Anthropology, Dibrugarh University.

Goswami, RG & Bhattacharyya, M 2015, 'Nutritional Status among adult Karbi Women of Kamrup District, Assam', *International Research Journal of Social Sciences,* Vol. 4, no. 9, pp. 17-20.

Haddad, L, Hoddinott, J & Alderman, H 1997, *Intrahousehold resource allocation in developing countries: models, methods and policy*, Johns Hopkins University Press, Baltimore.

Hsieh, SD & Muto, T 2004, 'A simple and practical index for assessing the risk of metabolic syndrome during the routine health check-ups', *Nippon Rinsho*, Vol. 62, pp. 1143-1149.

Islam Barbhuiya, AFG & Das, R 2013, 'Gender differences in nutritional status among the adult Meiteis of Cachar district of Assam, India,' *Eurasian Journal of Anthropology*, Vol. 4, no. 2, pp. 36"44.

James, WPT, Ferro-Luzzi, A & Waterlow, JC 1988, 'Definition of chronic energy deficiency in adults', *European Journal of Clinical Nutrition*, Vol. 42, no. 969–981.

James, WPT, Mascie-Taylor, CGN, Norgan, NG, Bristrian, BR, Shetty, P & Ferro-Luzzi, A 1994, 'The value of arm circumference measurements in assessing chronic energy deficiency in Third World adults', *European Journal of Clinical Nutrition*, Vol. 48, pp. 883-894.

Kishor, S & Neitzel, K 1996, *The status of women: indicators for twenty-five countries*, Demographic and Health Surveys, Comparative Studies No. 21, Macro International, Inc., Calverton, MD.

Kopelman, PG 2000, 'Obesity as a medical problem', *Nature*, Vol. 404, pp. 635-643.

Kulkarni, S & Sharma, ER 2001, Nutrition among Karbi tribe of Assam: an anthropometric assessment, *Journal of Indian Anthropological Society*, Vol. 36, pp. 223-229.

Lee, RD & Nieman, DC 2003, *Nutritional assessment*, McGraw Hill, New York.

Mungreiphy, NK & Kapoor, S 2010, 'Socio-economic changes as covariates of overweight and obesity among Tangkhul Naga tribal women of Manipur, North-East India', *Journal of Biosocial Science*, Vol. 42, pp. 289–305.

Naidu, AN 1994, 'Body mass index: a measure of the nutritional status in the Indian population, *European Journal Clinical Nutrition*, Vol. 48, no. 3, pp. 131–140.

National Family Health Survey-IV 2015-2016, *Health and family welfare*, Govt. of India.

Norgan, NG 1990, 'Body mass index and body energy stores in developing countries', *European Journal of Clinical Nutrition*, Vol. 44, pp. 79–84.

Sengupta, S & Hussain, US 2014, 'Age changes in somatometric characters among the adult Boro Kacharis of Assam: a preliminary appraisal', *Bulletin of the Department of Anthropology, Dibrugarh University*, Vol. 42, pp. 81-88.

Shetty, PS & James, WPT 1994, *Body mass index: a measure of chronic energy deficiency in adults*. Food and Nutrition Paper 56, Food and Agricultural Organization, Rome.

Sikdar, M & Ahmed Das, F 2006, 'Growth progression with ages in physical and physiological variables among the Bengali population of Dibrugarh, Assam', *Bulletin of the Department of Anthropology, Dibrugarh University*, Vol. 34, pp. 45-52.

Singh, MS 2013, *Obesity and self-reported morbidity in relation to lifestyle and diet among the Meitei of Manipur*, Ph.D. thesis, Department of Anthropology, North-Eastern Hill University.

Smith, LC 1998, 'Macroeconomic adjustment and the balance of bargaining power in rural West African households', Paper Presented at the *American Agricultural Economics Association Annual Meetings*, Salt Lake City, Utah.

Temsutola 2010, *Body composition and nutritional status among the Ao adults*, Ph.D. thesis, Department of Anthropology, Northeastern Hill University.

Ulijaszek, SJ & Kerr, DA 1999, 'Anthropometric measurement error and the assessment of nutritional status', *British Journal of Nutrition*, Vol. 82, pp. 165-177.

Webb, GP 2002, *Nutrition: a health promotion approach*, 2nd edn, Arnold Publishers, London.

Weiner, JS & Lourie, JA 1981, *Practical human biology,* Academic Press, London.

WHO Expert Consultation 2008, Waist circumference and waist-hip ratio.

World Health Organization 2000, *The Asia Pacific perspective, redefining obesity and its treatment*, International Diabetes Institute, Health Communications Australia Pvt. Ltd.

World Health Organization 2004, 'Appropriate body-mass index for Asian populations and its implications for policy and intervention strategies', *The Lancet*, Vol. 363, pp. 157-163.

Chapter - 7

Determinants of Overweight and Obesity among Meitei Males of Manipur

—Maibam Samson Singh

Introduction

Overweight and obesity is now considered as major health issues affecting people around the world. Obesity is simply defined as the excessive accumulation of fat in the body. Both overweight and obesity are often used interchangeably, but they do not represent the same situation. Obesity is a state of being seriously overweight to the stage where it becomes a risk factor for the degeneration of health. The individual shows differences in the amount and regional distribution of fat. The excess accumulation of fat in the abdominal region is considered one of the greatest risks for different diseases (WHO, 1998). The causes of obesity are complex and multi-factorial which involves the interaction between multiple genes and the environment. The change in the gene pool is not rapid; it is the environmental and social factors that undergo tremendous changed and caused the epidemic (Clement and Ferre, 2003). The obesogenic environment has attributed to the rapid increase in the prevalence of obesity across the world. A dominant explanatory framework for the emergence of the obesogenic environment is that of nutrition transition (Popkin, 2004). Central to this transition is rapidly changing lifestyles through globalization, urbanization and technological development. World Health Organization reports an estimated 200 million obese adults worldwide by 1995, which was increased to over 300 million by 2000. In 2010, the prevalence of adults overweight is more than 1.4 billion and more than 400 million adults are obese (WHO, 2011). More than 1.9 billion adults were overweight and an estimated 600 million were obese in 2014 (WHO, 2016). The Global Health Observatory data shows that 2.8 million people die every year as a result of being overweight and obese (WHO, 2016). According to NFHS-3 (IIPS, 2007), among the

Indian state, the prevalence of overweight among men was reported highest in Punjab (22.2%) followed by Kerala (17.8%) and Delhi (16.8%). In Northeast India, the prevalence of overweight among men was highest in Sikkim (11.9%). Obesity increases with age and it increases most rapidly when people are in their twenties and early thirties and continues to rise until they reach their late fifties (Rolland-Cachera et al., 1991; Flegal et al., 1998). Individuals under thirty years old, who have been consistently thin, are still at risk of developing obesity once they are in their thirties (Bulik and Allison, 2001). Longitudinal studies also show that people generally gain weight as they become older (Guo et al., 1999; Heitmann and Garby, 1999). In developing countries like India, there occurs an increase in body weight with aging (Agrawal, 2002).

The positive relationship between economic conditions and obesity are reported in many countries like Brazil (Montiero et al., 2004), Cameroon (Fezeu et al., 2006), India (Reddy et al., 2002) Jordan (Montiero et al., 2004) and Madagascar (Monteiro et al., 2004). Despite this positive relationship, the burden of obesity shift towards the poor with increasing economic and social development (Monteiro et al., 2004; Song, 2006). Overweight and obesity are now dramatically increased in low and middle-income countries (Brewis, 2011). Occupational physical activity is an important determinant of weight gain. (Keighley et al., 2006) found that adults in American Samoa who were engaged in farm work have lower BMIs than those not engaged in such work. An inverse association between obesity and educational levels has been reported in numerous affluent populations (Sobal and Stunkard, 1989; Stam-Moraga et al., 1999; Wardle and Griffith, 2001). Molarius et al., (2000) found that lower education is associated with higher BMI in almost all of the WHO MONICA populations. A Swedish study also reported the highest BMI among the lowest level of education in both men and women (Sundquist and Johansson, 1998). However, there are also figures that show an increase in the prevalence of obesity during the last decade of the 20th century among high-educated men and women (Visscher et al., 2002). A positive association between obesity and educational level has been shown in some eastern and central European countries (Molarius et al., 2000). Physical activity helps in maintaining healthy weight as it influenced body composition, metabolism, and increasing expenditure (Nowicka and Flodmark, 2007). There has been a great decline in occupationally related activity since the turn of the twentieth century (Popkin et al., 2005). Low levels of physical activity are associated with an increased risk of obesity (Erlichman et al.,2002). Jobs requiring heavy manual labour have been largely replaced by jobs in service and high technology sectors, which require minimal physical exertion in developing countries (French et al., 2001). The increased use of automobiles and public transportation systems along with increasing time spent on watching television, playing electronic games and using computers encourages sedentary behaviours (Hill and Peters, 1998; Brownell, 2002). Studies indicated that short sleep is associated with an increased risk of obesity in many countries in the US, Australia and Europe (Cappuccio et al., 2008; Magee et al., 2010b). Modest reductions in sleep duration are associated with significant increases in obesity risk among adults (Gangwisch et al., 2005; Vorona et al., 2005). Studies in Japanese children of the Toyama Birth Cohort also show an association between shorter sleep duration and increase overweight (Sekine et al., 2002; Sugimori et al., 2004).

Obesity has become an epidemic worldwide. The prevalence of both overweight and obesity is now increasing rapidly in many developing countries like India through socio-economic development, nutritional transition as well as increasing urbanization. Manipur, one of the Northeastern states of India has undergone tremendous change in the past few years in terms of urbanization and socio-economic development. Such a transition has resulted in changes in terms of occupation, economy, dietary intake as well as overall lifestyles. However, our understanding of the causes and consequences of obesity remains poor in the state due to limited available data. Therefore, the present study has been conducted to examine the prevalence of overweight in relation to socio-economic conditions and certain lifestyle factors among Meitei males of Manipur.

Materials and Methods

The present research is conducted among adult Meitei males and adolescent boys residing in the urban areas of Lamphel subdivision, Imphal West district of Manipur. Manipur is a small state located on the far Northeastern edge of the map of India. The state is bound by Cachar and North Cachar districts of Assam on the west, Kohima and Phek districts of Nagaland on the north, and Aizawl district of Mizoram on the southeast. Its entire eastern border is shared with Myanmar. It is, therefore, the gateway of India to Southeast Asia through the Moreh town of Manipur passing through National Highway 39 (Indo Burma Road). About 10 percent of the Manipur's area is constituted by plains while the rest is covered by undulating hills. Most of the plain area is covered by the Imphal valley. The State comprises of three major ethnic groups, namely Meitei, Nagas, and Kuki-Chins-Mizo. The Meitei, who inhabit the valley, are numerically the most predominant community in the state. At present, several people from the hills have also migrated and settled in the valley. Besides these three major ethnic groups, small Muslim population locally known as Meitei Pangal are settled in Manipur, mainly in the valley areas. A small percentage of other groups like Bengalis, Nepalese, Jain, and Sikhs who have migrated from time to time are also residing in different areas of Manipur. Agriculture is the main occupation of the people of Manipur. It has an important place in the economy of the state. The agricultural sector contributes a major share to the total state domestic place and provides employment to about 52.2 percent of the total workforce in Manipur. Urbanization has accelerated in the past few years which bring tremendous changes to the State.

The data is collected from the Meitei adolescent boy (aged 14 to 19 years) and adult men (aged 20 to 60 years) living in the urban areas of Lamphel subdivision of Imphal west district. The Imphal West district is the most developed and urbanized districts of Manipur. Imphal, the State capital is the nodal functional center of this district. Being the main capital, Imphal west is the district where all the main important offices, institutions, hotels and markets are located. A sample size of 256 adolescent boys and 350 adult men (total of 606 subjects) was collected randomly following house to house visit. Height and body weight were measured using an anthropometric rod and a weighing scale to the nearest of 0.1cm and 0.5kg respectively with subjects wearing light clothes following a standard technique (Lohman et al., 1998).

Data on household income were collected directly from the subjects or the head of the households. The incomes were cross-checked against certain aspects of socio-economic conditions like housing condition, types of occupation, landholding and monthly expenditure. The per capita monthly income of the households was classified as a high-income group (above 75th percentile); middle-income group (50th to 75th percentile) and low-income group (below 50th percentile). Data on educational attainment of individuals were classified into five categories, namely, illiterate, primary, secondary, higher secondary and graduate and above. The illiterate in the present study are those individuals who were not able to read or write. Since the illiterate subjects are very negligible in number, therefore they are clubbed with the category of primary education. The primary level of education includes those individuals who studied up to class V. In the secondary level of education, we included those individuals who attended standard VI to X. Higher secondary level of education included those who attended standard XI to XII. In the graduate and above level of education, we included those individuals who have completed graduation and those pursuing higher studies. Data on the occupation of each subject were classified into five categories such as government employee, self-employed, business, student, and no occupation. Information on lifestyles such as physical activity, sleeping hours was collected from each subject following a recalled method of two-week periods. The data on physical activity were divided into two categories: less than one hour and one hour and above. The sleeping hours were classified into three groups, namely six hours, seven to eight hours and more than eight hours.

The data were analyzed using both MS-Excel software and SPSS for the present research. The parameters taken were analyzed statistically to find out the mean, standard error of the mean for the anthropometric measurements. Prevalence of overweight and underweight has been evaluated using the WHO International value (WHO, 2000). According to WHO classification, BMI $\geq$ 30kg/m^2 has been considered as obese and BMI between 25.0-29.9kg/m^2, considered as overweight. BMI between 18.5-24.9kg/m^2 indicates normal, and BMI below 18.5kg/m^2 indicates underweight. Prevalence of overweight and underweight was calculated in relation to different socio-economic conditions and lifestyle factors. In order to test the level of significance, both t-test and chi-square have been used in the present study. Regression analysis has also been carried out to find the risks of overweight and obesity.

Results

The basic data on mean height, weight, and BMI along with standard error of Meitei adolescent boys and adult men are given in Table 7.1. The mean height was higher among adult men (165.1±0.3) than the adolescent boys (161.3±0.5). The differences in the mean height among the adolescent boys and adult men are statistically significant ($p<0.01$). Similarly, the mean body weight among the adult men (63.1± 0.5) was significantly ($p<0.01$)

Table 7.1: Height, weight, and BMI among Meitei adolescent boy and adult men

Age group (year)	*N*	*Height (cm) ±SE*	*Weight (kg) ±SD*	*BMI±SE*
14-19	256	161.3±0.5	53.8±0.4	20.6±0.2
20-60	350	165.1±0.3**	63.1± 0.5**	23.1±0.2**

N=sample size; SE= standard error; **=0.01

higher than the adolescent boys (53.8±0.4). The table further shows that the mean BMI value among the adult men (23.1±0.2) was significantly ($p<0.01$) higher than the adolescent boys (20.6±0.2).

According to WHO International value (Table 7.2), the prevalence of overweight (19.4%) was higher among the adult men compared to the adolescent boys (3.1%). The frequency of underweight was higher among adolescent boys (8.6%) than adult men (2.3%). The distributions were statistically significant ($p<0.001$). The overall prevalence of underweight and overweight among Meitei was 5.0 percent and 12.5 percent respectively.

Table 7.2: Prevalence of overweight and obesity among Meitei adolescent boys and adult men

Age group (year)	*N*	*Underweight*	*Normal*	*Overweight and obesity*
14-19	256	22(8.6%)	226(88.3%)	8(3.1%)
20-60	350	8(2.3%)	274(78.3%)	68(19.4%) ***
Total	606	30(5.0%)	500(82.5%)	76(12.5%)

***=0.001

The prevalence of overweight and underweight among adolescent boys and adult men in relation to family income are given in Table 7.3. The frequency of overweight (15.1%) was found only among the adolescent boys belonging to the high-income families. Whereas, the prevalence of underweight (16.8%) was found among adolescent boys belongs to low-income families. The differences in the prevalence were found statistically significant ($p<0.001$). Among the adult men, the frequency of overweight (34.7%) was higher in high-income families. This was followed by the frequency of overweight among the middle-income families (18.8%) and low-income families (8.2%). The frequency of underweight (6.6%) among adult men was found only in low-income families. The differences were significant at $p<0.001$. The overall prevalence of overweight is increased along with the increase in the levels of family incomes (Table 7.3.1). The total frequency of overweight (27.7%) was higher among men belongs to high-income families followed by middle-income families (12.2%) and low-income families (4.0%). The frequency of underweight (11.9%) was found among men in low-income families. The differences in the prevalence of underweight and overweight were significant statistically at $p<0.001$.

Table 7.3: Prevalence of overweight and obesity in relation to family income among Meitei adolescent boys and adult men

Age group (year)	*Income*	*N*	*Underweight*	*Normal*	*Overweight % of obesity*
14-19	HIG	53	0	45(84.9%)	8(15.1%)
	MIG	72	0	72(100.0%)	0
	LIG	131	22(16.8%)	109(83.2%)	0 ***
Total		256			
20-60	HIG	95	0	62(65.2%)	33(34.7%)
	MIG	133	0	108(81.2%)	25(18.8%)
	LIG	122	8(6.6%)	104(85.2%)	10(8.2%) ***
Total		350			

HIG: high income; MIG: middle income; LIG: lower income; ***=0.001

Table 7.3.1: Overall prevalence of overweight and obesity in relation to family income among Meitei adolescent boys and adult men

Age group (year)	*Income*	*N*	*Underweight*	*Normal*	*Overweight and obesity*
14-60	HIG	148	0	107(72.3%)	41(27.7%)
	MIG	205	0	180(87.8%)	25(12.2%)
	LIG	253	30(11.9%)	213(84.1%)	10(4.0%) ***
Total		606			

HIG: high income; MIG: middle income; LIG: lower income; ***=0.001

The occurrence of overweight (3.1%) was found the same among the adolescent boys in both the secondary and the higher secondary education (Table 7.4). The frequency of underweight was higher among adolescent boys who attained secondary education (12.4%) than higher secondary education (4.7%). Among the adult men, the frequency of overweight (33.3%) was higher in primary education. This was followed by overweight among adult men who attained graduate and above (19.3%), higher secondary (18.8%) and secondary education (14.3%). The prevalence of underweight was higher among men who attained secondary education (28.6%) followed by higher secondary education (2.6%) and graduate and above education (1.4%). The differences in the prevalence of underweight and overweight were statistically significant ($p<0.01$). The overall prevalence of overweight and underweight in relation to educational levels is given in Table 7.4.1. The overall frequency of overweight (33.3%) among men was recorded higher in primary education. The graduate (19.3%), higher secondary (10.6%) and secondary (3.7%) followed the higher frequency of overweight. The frequency of underweight was higher in secondary education (13.2%) followed by higher secondary (3.7%) and graduate and above (1.4%). The differences were significant (p<0.001).

Table 7.4: Prevalence of overweight and obesity in relation to the educational level among Meitei adolescent boys and adult men

Age group (year)	*Education*	*N*	*Underweight*	*Normal*	*Overweight and obesity*
14-19	Secondary	129	16(12.4%)	109(84.5%)	4(3.1%)
	Hr. Secondary	127	6(4.7%)	117(92.2%)	4(3.1%)
Total		256			
20-60	Primary	9	0	6(66.7%)	3(33.3%)
	Secondary	7	2(28.6%)	4(57.1%)	1(14.3%)
	Hr. Secondary	117	3(2.6%)	92(78.6%)	22(18.8%)
	Graduate+	217	3(1.4%)	172(79.3%)	42(19.3%) **
Total		350			

Table 7.4.1: Overall prevalence of overweight and obesity in relation to the educational level among Meitei adolescent boys and adult men

Age group (year)	*Education*	*N*	*Underweight*	*Normal*	*Overweight % of obesity*
14-60	Primary	9	0	6(66.7%)	3(33.3%)
	Secondary	136	18(13.2%)	113(83.1%)	5(3.7%)
	Hr. Secondary	244	9(3.7%)	209(85.7%)	26(10.6%)
	Graduate+	217	3(1.4%)	172(79.3%)	42(19.3%) ***
Total		606			

***=0.001

According to WHO International value (Table 7.5), the prevalence of overweight (3.2%) and underweight (8.9%) was found only in the students among the adolescent boys. Among the adult men, overweight (38.4%) was higher in the government employees. Higher frequency of overweight among the government employees was followed by overweight among the businessmen (13.5%), no occupation (10.0%), students (7.2%) and self-employed (2.8%). The higher frequency of underweight (4.8%) was found among student. The prevalence of underweight was found more or less the same between businessmen (2.9%) and self-employs (2.8%). The differences in the prevalence of overweight and underweight in relation to occupation were found significant (p<0.001). The overall prevalence of overweight and underweight in relation to occupation is given in Table 7.5.1. The frequency of overweight (38.4%) was found higher among government employees. These were followed by overweight among businessmen

(13.5%), no occupation (10.0%), students (3.8%) and self-employed (2.3%). The prevalence of underweight was higher among student (8.3%). The frequency of underweight among government employee, businessmen and self-employ was 0.9 percent, 2.9 percent, and 2.3 percent respectively. The differences in the prevalence of overweight and underweight were statistically significant (p<0.001).

Table 7.5: Prevalence of overweight and obesity in relation to occupation among Meitei adolescent boys and adult men

Age group (year)	*Occupation*	*N*	*Underweight*	*Normal*	*Overweight and obesity*
14-19	Self-employ	8	0	8(100.0%)	0
	Student	248	22(8.9%)	218(87.9%)	8(3.2%)
Total		256			
20-60	Govt. employee	112	1(0.9%)	68(60.7%)	43(38.4%)
	Business	140	4(2.9%)	117(83.6%)	19(13.5%)
	Self-employ	36	1(2.8%)	34(94.4%)	1(2.8%)
	Student	42	2(4.8%)	37(88.0%)	3(7.2%)
	No	20	0	18(90.0%)	2(10.0%) ***
Total		350			

***=0.001

Table 7.5.1: Overall prevalence of overweight and obesity in relation to occupation among Meitei adolescent boys and adult men

Age group (year)	*Occupation*	*N*	*Underweight*	*Normal*	*Overweight % of obesity*
14-60	Govt. employee	112	1(0.9%)	68(60.7%)	43(38.4%)
	Business	140	4(2.9%)	117(83.6%)	19(13.5%)
	Self-employ	44	1(2.3%)	42(95.4%)	1(2.3%)
	Student	290	24(8.3%)	255(87.9%)	11(3.8%)
	No	20	0(0.0%)	18(90.0%)	2(10.0%) ***
Total		606			

***=0.001

According to WHO International value (Table 7.6), the prevalence of overweight (33.3%) was found significantly (p<0.001) higher among the adolescent boys who exercised below one hour than over one hour (1.6%). The frequency of underweight (9.0%) was found among adolescent boys who exercised one hour and above. Among the adult men, the frequency of overweight (19.9%) was recorded higher among those subjects who exercised below one hour. Overweight (10.5%) was observed among the adult men who exercised over one hour. The result further showed that underweight (2.4%) was found among adult men exercised below one hour.

The overall prevalence of overweight (20.4%) was found higher among men who exercised below one hour (Table 7.6.1). The frequency of overweight (2.3%) was found low among men who exercised over one hour. However, the overall frequency of underweight was higher among men exercised one hour above (8.4%) than exercised below one hour (2.3%). The differences in the prevalence of overweight and underweight were significant statistically ($p<0.001$).

Table 7.6: Prevalence of overweight and obesity in relation to physical activity among Meitei adolescent boys and adult men

Age group (year)	*Physical activity*	*N*	*Underweight*	*Normal*	*Overweight and obesity*
14-19	<1hour	12	0	8(66.7%)	4(33.3%)
	=1hour	244	22(9.0%)	218(89.4%)	4(1.6%) ***
Total		256			
20-60	< 1hour	331	8(2.4%)	257(77.6%)	66(19.9%)
	=1hour	19	0	17(89.5%)	2(10.5%)
Total		350			

***=0.001

Table 7.6.1: Overall prevalence of overweight and obesity in relation to physical activity among Meitei adolescent boys and adult men

Age group (year)	*Physical activity*	*N*	*Underweight*	*Normal*	*Overweight and obesity*
14-60	<1hour	343	8(2.3%)	265(77.3%)	70(20.4%)
	≥1hour	263	22(8.4%)	235(89.3%)	6(2.3%) ***
Total		606			

***=0.001

Table 7.7 shows that the frequency of overweight (6.3%) was higher among the adolescent boys sleeping over eight hours followed by overweight (3.1%) among seven-eight hours sleeping time. The frequency of underweight was also higher among adolescent boys sleeping over eight hours (12.4%). The prevalence of underweight among adolescent boys sleeping between seven to eight hours and six hours was 8.6 percent and 3.7 percent respectively. Among the adult men, overweight (23.9%) was found among those sleeping six hours. The frequency of overweight among men sleeping seven to eight hours and above eight hours was 18.9 percent and 11.1 percent respectively. The frequency of underweight (2.7%) was found among men sleeping seven to eight hours. The overall prevalence of overweight in association with sleeping hours is given in Table 7.7.1. The frequency of overweight (15.1%) was found higher among men sleeping six hours. This was followed by overweight among men sleeping seven to eight hours (12.6%) and over eight hours (7.3%). The table further shows that the frequency of underweight was higher among men sleeping over eight hours (9.8%) followed by seven to eight hours (5.1%) and six hours (1.4%).

Table 7.7: Prevalence of overweight and obesity in relation to sleeping duration among Meitei adolescent boys and adult men

Age group (year)	*Sleeping duration*	*N*	*Underweight*	*Normal*	*Overweight and obesity*
14-19	6 hour	27	1(3.7%)	26(96.3%)	0
	7-8 hour	197	17(8.6%)	174(88.3%)	6(3.1%)
	above 8 hr.	32	4(12.4%)	26(81.3%)	2(6.3%)
Total		256			
20-60	6 hour	46	0	35(76.1%)	11(23.9%)
	7-8 hour	295	8(2.7%)	231(78.3%)	56(18.9%)
	above 8 hr.	9	0(0.0%)	8(88.9%)	1(11.1%)
Total		350			

Table 7.7.1: Overall overweight and obesity in relation to sleeping duration among Meitei adolescent boys and adult men

Age group (year)	*Sleeping duration*	*N*	*Underweight*	*Normal*	*Overweight and obesity*
14-60	6 hour	73	1(1.4%)	61(83.5%)	11(15.1%)
	7-8 hour	492	25(5.1%)	405(82.3%)	62(12.6%)
	above 8 hr.	41	4(9.8%)	34(82.9%)	3(7.3%)
Total		606			

The risk of overweight and obesity among men is given in Table 7.8. The adult men (19.5%) show increased risks (OR-7.475; Cl-3.523-15.860) of overweight compared to the adolescent boys (3.1%). The table further reveal that high-income families (27.7%) showed a higher prevalence of overweight. This was followed by middle income families (12.2%) and lower-income families (4.0%). The result further indicated that the risk of overweight showed to be higher among men who attained primary education (33.3%) and government employees (38.4%). The exercise below one hour (20.4%) and six hours sleeping duration (15.1%) were also observed the higher risk of overweight and obesity among men in the present study.

Discussion

Obesity is relatively new in human evolutionary history. Until the start of the Neolithic age some 10,000 years ago when farming lifestyles became prevalent, diets were based on foods caught and gathered, such as lean

Table 7.8: Risks of overweight and obesity among Meitei men by individuals and family characteristics

Variables	*Total N (%)*	*Overweight and obesity (%)*	*OR (95% cl)*	*P*
Age group				
14-19	256(42.2%)	8(3.1%)	1	
20-60	350(16.8%)	68(19.5%)	7.475(3.523-15.860)	0.001
Income				
LIG	253(41.8%)	10(4.0%)	1	
MIG	205(33.8%)	25(12.2%)	3.375(1.581-7.207)	0.002
HIG	148(24.4%)	41(27.7%)	9.311(4.497-19.277)	0.001
Education				
Primary	9(1.5%)	3(33.3%)	1	
Secondary	136(22.4%)	5(3.7%)	0.076(0.015-0.397)	0.002
Hr. secondary	244(40.3%)	26(10.6%)	0.239(0.056-1.011)	0.052
Graduate+	217(35.8%)	42(19.3%)	0.480(0.115-1.998)	0.313
Occupation				
Govt. employee	112(18.5%)	43(38.4%)	1	
Business	140(23.1%)	19(13.5%)	0.252(0.136-0.466)	0.001
Self-employ	44(7.3%)	1(2.3%)	0.037(0.005-0.281)	0.001
Student	290(47.8%)	11(3.8%)	0.063(0.031-0.129)	0.001
No	20(3.3%)	2(10.0%)	0.178(0.039-0.807)	0.025
Physical activity/exercise				
=1hour	263(43.4%)	6(2.3%)	1	
<1hour	343(56.6%)	70(20.4%)	10.983(4.690-25.721)	0.001
Sleeping hour				
6 hours	73(12.0%)	11(15.1%)	1	
7-8 hours	492(81.2%)	62(12.6%)	0.813(0.406-1.627)	0.558
Above 8 hours	41(6.8%)	3(7.3%)	0.445(0.117-1.698)	0.236

OR-odd ratio; cl-confidence level; 1-reference value.

meat, nuts, and fibrous plants (Brewis, 2011). The only readily available drink was water. Since walking was the only way to go around, people naturally exercised on a daily basis. The possibilities of being overweight and obesity in such an environment is rare. However, the new millennium has signaled an important transition for our species with more people becoming overweight and obese (Mendez et al., 2005; Brewis, 2011). In several of the developing nations across the world, including Africa, a continent usually associated with incidences of starvation and widespread malnutrition, the prevalence of overweight and obesity is increasing (Berman, 2009; Brewis, 2011). Currently, the most rapid increases in adult overweight and obesity are occurring in middle-income countries like China and India (Monteiro et al., 2004). This reflects the speediness of socio-economic development, nutrition transition, urbanization and modernization (Sobal and Stunkard, 1989; Brownell, 2002; Popkin, 2004).

Manipur, a small state in Northeast India has begun to experience tremendous changes in the last decade. The socio-economic status of many of its inhabitants has improved significantly through urbanization. There is tremendously increased in vehicular and motor transport along with an increasing number of fast food restaurants and roadside vendors. Such a transition has, over the years, resulted in changes among the Meitei population in terms of occupation, economy, food consumption, physical activity levels, and overall lifestyles. Prevalence of overweight among Meitei males is differed by age. Adult men show a higher frequency of overweight than adolescent boys. In general, elderly people are relatively less involved in physical activities and more in sedentary lifestyles. With increasing age, there is a progressive increase in fat and decrease in fat-free mass (Molarious et al., 2000). Further, the patterns of overweight differ across different socio-economic groups. The higher prevalence of overweight among high-income families in the present study is consistent with studies from developing countries (Monteiro et al., 2004). The prevalence of overweight increases with increasing economic conditions of the families. The pattern of obesity appears to be different with regard to educational attainment. Prevalence of overweight is found the same between secondary and higher secondary education among adolescent boys. However, among adult, the frequency of overweight was higher in primary education. Overall result further show that overweight shows a gradual increase along with the increase in the levels of education from the secondary schooling onwards. Socio-economic conditions especially economic conditions can influence people's dietary pattern, physical activity, and overall lifestyles. Rapid economic and social development has caused rapid changes in dietary pattern, physical activity, overall lifestyles (Popkin and Gordon-Larson, 2004). Further, rich and wealthy persons are more likely to be obese especially in developing countries as they have enough resources at their disposal to eat well all the time (Sobal and Stunkard, 1989; Brewis, 2011). On the other hand, motorized transportation and labour saving household devices introduced by technological and socio-economic development resulted in a marked reduction of physical activities (Song, 2006). Such a transition could have a major impact on the rich and wealthy families as they can afford it more readily than poor.

The frequency of overweight is significantly higher among government employees. This was followed by a higher frequency of overweight among businessmen. Meitei men who are self-employed show the least prevalence of overweight. Government employees may relatively involve in less physical activity and more sedentary life than other occupations. The decrease in occupational related activity coupled with increasingly sedentary lifestyles could result in an increase in mean weight (Brownson et al., 2005). With the ever-increasing urbanization and technological development, most of the occupational work can now be performed with minimum physical activities (Mondal et al., 2008). On the other hand, diet and other personal lifestyle behaviours that influence body weight such as walking for transport, alcohol consumption and smoking are rarely considered to be related to occupation (Allman-Farinelli et al., 2010).

The negative association between overweight and physical activity in both the adolescent boys and adult men in the present study is consistent with other findings (Atlantis et al., 2006). Inactive physical activity and more sedentary lifestyles may lead to unhealthy weight gains due to the accumulation of unwanted fat in the body. However physical activities, even if slightly below the recommended levels, can effectively reduce body fat (Atlantis et al., 2006). The built environment and political situation are of particular interest in decreasing physical activities among the Meitei population. Increasing land use for development, traffic density, restricted sports grounds, and parks provide restrictions on physical activity promoting behaviour (Black and Macinko, 2008). This may relatively decrease in physical activity by restricting people to the confines of the home environment leading to more sedentary lifestyles, which in turn, lead to the development of obesity in due course of time. The association between sleeping hours and the prevalence of overweight is controversial. Recent studies have found that short sleep is associated with an elevated risk of obesity (Cappuccio et al., 2008; Patel and Hu, 2008) whereas others found no evidence of an association between sleep duration and obesity (Gangwisch et al., 2005; Gottlieb et al., 2006). The association between obesity and sleeping time shows mixed results among adolescents and adults in the present study. However, the overall distribution of overweight is found to be higher among those who sleep less hour. The study further indicates the risks of overweight and obesity among those who sleep relatively short hour. Restricted or short sleep alters secretion of hormone leptin and ghrelin that play an important role in the regulation of hunger and appetite (Spiegel et al., 2009; Magee et al., 2010b). With the increase in general appetite and food intake, short sleep could lead to the selection of calorie-dense foods in particular and increased in fat intake leading to weight gain (Taheri, 2006; Shi et al., 2008). Sleep deprivation results in tiredness, fatigue and daytime sleep; this could contribute to reduced daytime activity and energy expenditure (Taheri, 2006). There is also the possibility that overall energy expenditure is less among short sleepers than the normal sleepers due to daytime sleepiness and decrease physical activity (Dinges et al., 1997)

In conclusion, the patterns of overweight and obesity are differed by age with adult men being more prone to become overweight suggests increasing risks of overweight with advancing age. Socio-economic conditions and certain lifestyle factors have been identified as an important determinant for explaining the prevalence of overweight and obesity in the present research. The increasing urbanization and socio-economic improvement coupled with technological development have predisposed many individuals to lifestyle factors associated with obesity. There is increased the availability of calorie dense foods, which, in turn, altered the traditional food environment. The altering food environment in Manipur could be observed with the increasing numbers of fast food restaurants and roadside vendors. The presence of available and affordable transportation systems and remote-controlled appliances have reduced physical activities. The technological development such as computer and video games exposed children to more sedentary lifestyles. Such a rapid changing lifestyle through urbanization and economic growth might lead to the increasing prevalence of obesity, which in turn lead to a challenging health problem in the near future. Therefore, there is a need to educate the public and create an awareness programme about the problem of obesity in Manipur. Health education programs and effective policies are urgently required to promote healthy eating and physical activity during the early childhood period. Regular monitoring of BMI and fat distribution for adults would be useful to control even modest weight gain with advancing age.

References

Agrawal, PK 2002, *Emerging obesity in the Northern Indian States: A serious threat for health*. Paper presented at the IUSSP Regional Conference, Bangkok, June 10-13.

Allman-Farinelli, MA, Chey, T, Merom, D and Bauman, AE 2010, 'Occupational risk of overweight and obesity: an analysis of the Australian Health Survey'. *Journal of Occupational Medicine and Toxicology*, Vol. 5, pp14-22.

Atlantis, E, Barnes, EH and Singh, MA 2006, 'Efficacy of exercise for treating overweight in children and adolescents: a systematic review'. *International Journal of Obesity*, Vol. 30, no. 7, pp. 1027-1040.

Berman, M 2009, *Africa-yes, Africa-has an obesity problem*. Newsweek, http://www.newsweek.com/id/212556.

Black, JL and Macinko, J 2008, 'Neighbourhoods and obesity'. *Nutrition Review*, Vol. 6, pp. 2-20.

Brewis, AA 2011, *Obesity: cultural and biocultural perspectives.* Rutgers University Press, New Brunswick, New Jersey.

Brownell, KD 2002, *The environment and obesity. In eating disorders and obesity: A comprehensive handbook*, ed. CG Fairburn, KD Brownell. 433–438. New York: Guilford Press. 2nd edition.

Brownson, RC, Boehmer, TK and Luke, DA 2005, 'Declining rates of physical activity in the United States: what are the contributors'? *Annual Review of Public Health*, Vol. 26, pp. 421-443.

Bulik, CM and Allison, DB 2001, 'The genetic epidemiology of thinness'. *Obesity Reviews*, Vol. 2, pp. 107-115.

Cappuccio, FP, Taggart, FM, Kandala, NB, Currie, A, Peile, E, Stranges, S and Miller MA. 2008, 'Meta-analysis of short sleep duration and obesity in children and adults'. *Sleep*, Vol. 31, pp. 610–626.

Clement, K and Ferre, P 2003, 'Genetics and pathophysiology of obesity'. *Pediatric Research*, Vol. 53, pp. 721-725.

Dinges, DF, Pack, F and Williams, K 1997, 'Cumulative sleepiness, mood disturbance, and psychomotor vigilance performance decrements during a week of sleep restricted to 4–5 hours per night'. *Sleep,* Vol. 20, pp. 267-267.

Ehrlichman, J, Kerbe, AL and James, WPT 2002, 'Physical activity and its impact on health outcomes. Paper 2: Prevention of unhealthy weight gain and obesity by physical activity: an analysis of the evidence'. *Obesity Review,* Vol. 3, pp. 273-287.

Fezeu, L, Minkoulou, E, Balkau, B, Kengne, AP, Awah, P, Unwin, N, Alberti, GKMM and Mbanya, JC 2006, 'Association between socio-economic status and adiposity in urban Cameroon'. *International Journal of Epidemiology*, Vol. 35, no.1, pp. 105-111.

Flegal, KM, Carroll, MD, Kuczmarski, RJ and Johnson, CL 1998, 'Overweight and obesity in the United States: prevalence and trends, 1960-1994'. *International Journal of Obesity*, Vol. 22, pp. 39-47.

French, SA, Story, M and Jeffery, RW 2001, 'Environmental influences on eating and physical activity. *Annual Review of Public Health*, Vol. 22, pp. 309-335.

Gangwisch, JE, Malaspina, D, Boden-Albala, B and Heymsfield, SB 2005, 'Inadequate sleep as a risk factor for obesity: analyses of the NHANES I'. *Sleep*, Vol. 28, pp. 1289-1296.

Gottlieb, DJ, Redline, S, Nieto, FJ, Baldwin, CM, Newman, AB and Resnick, HE 2006, 'Association of usual sleep duration with hypertension: the sleep heart health study'. *Sleep,* Vol. 29, pp. 1009-1014.

Guo, SS, Zeller, C, Chumlea, WC and Siervogel, RM 1999, 'Aging, body composition, and lifestyle: the Fels longitudinal study'. *American Journal of Clinical Nutrition*, vol. 70, pp. 405-411.

Heitmann, BL, and Garby, L 1999, 'Patterns of long-term weight changes in overweight developing Danish men and women aged between 30 and 60 years'. *International Journal of Obesity Related Metabolic Disorder*, Vol. 23, pp. 1074-1078.

Hill, JO and Peters, JC 1998, 'Environmental contributions to the obesity epidemic'. *Science*, Vol. 280, pp. 1371-1374.

IIPS 2007, *National Family Health Survey (NFHS-3), 2005–2006, India*. Mumbai: International Institute for Population Sciences.

Keighley, ED, McGarvey, ST, Turituri, P and Vial, S 2006, 'Farming and adiposity in Samoan adults'. *American Journal of Human Biology, Vol.* 18, pp. 112-122.

Lohman, TG, Roch, AF, and Martorell, R 1998, *Anthropometric standardization reference manual*, Human Kinetics Books, Chicago, USA.

Magee, CA, Iverson, DC and Caputi, P 2010b, 'Sleep duration and obesity in middle-aged Australian adults'. *Obesity*, Vol. 18, pp. 420-421.

Mendez, MA, Monteiro, CA and Popkin, BM 2005, 'Overweight exceeds underweight among women in most developing countries'. *American Journal of Clinical Nutrition*, Vol. 81, no. 3, pp. 714-721.

Molarius, A, Seidell, JC, Sans, S, Tuomilehto, J and Kuulasmaa, K 2000, 'Educational level, Relative body weight, and changes in their association over 10 Years: An international perspective from the WHO MONICA Project'. *American Journal of Public Health,* Vol. 90, pp. 1260-1268.

Monda, KL, Adair, LS, Zhai, F and Popkin, BM 2008, 'Longitudinal relationships between occupational and domestic physical activity patterns and body weight in China'. *European Journal of Clinical Nutrition*, Vol. 62, pp. 1318-1325.

Monteiro, CA, Moura, EC, Conde, WL and Popkin, BM 2004, 'Socio-economic status and obesity in adult populations of developing countries: a review'. *Bulletin of World Health Organization,* Vol. 82, pp. 940-946.

Nowicka, P and Flodmark, CE 2007, 'Physical activity-key issues in the treatment of childhood obesity'. *Acta Paediatrica,* Vol. 96, no.454, pp. 39-45.

Patel, SR and Hu, FB 2008, 'Short sleep duration and weight gain: A systematic review'. *Obesity*, Vol. 16, pp. 643-653.

Popkin, BM, Duffey, K and Gorden-Larsen, P 2005, 'Environmental infuences on food choice, physical activity, and energy balance'. *Physiological Behaviour*, Vol. 86, pp. 603-613.

Popkin, BM and Gordon-Larsen, P 2004, 'The nutrition transition: worldwide obesity dynamics and their determinants'. *International Journal of Obesity*, Vol. 28, pp. 2S-9S.

Popkin, BM 2004, 'The nutrition transition: An overview of world patterns of change'. *Nutrition Reviews, Vol.* 62, pp. 140S-143S.

Reddy, KKR, Rao, AP and Reddy, TPK 2002, 'Socio-economic status and the prevalence of coronary heart disease risk factors'. *Asia Pacific Journal of Clinical Nutrition*, Vol. 11, no. 2, pp. 98-103.

Rolland-Cachera, MF, Cole, TJ, Sempe, M, Tichet, J, Rossignol, C and Charraud, A 1991, 'Body mass index variations: centiles from birth to 87 years'. *European Journal of Clinical Nutrition*, Vol. 45, pp. 13-21.

Sekine, M, Yamagami, T and Handa, K 2002, 'A dose-response relationship between short sleeping hours and childhood obesity: results of the Toyama Birth Cohort Study'. *Child Care Health Development*, Vol. 28, pp. 163-170.

Shi, Z, McEvoy, M, Luu, J and Attia, J 2008, 'Dietary fat and sleep duration in Chinese men and women'. *International Journal of Obesity*, Vol. 32, pp. 1835-1840.

Sobal, J, and Stunkard, AJ 1989, 'Socio-economic status and obesity: A review of the literature'. *Psychological Bulletin,* Vol. 105, pp. 260-275.

Song, YM 2006, 'Commentary: Varying relation of socio-economic status with obesity between countries at different stages of development'. *International Journal of Epidemiology,* Vol. 35, no. 1, pp. 112-113.

Spiegel, K, Tasali, E, Leproult, R and Van Cauter, E 2009, 'Effects of poor and short sleep on glucose metabolism and obesity risk'. *Nature Reviews Endocrinology,* Vol. 5, pp. 253-261.

Stam-Moraga, MC, Kolanowski, J, Dramaix, M, deBacker, G, Kornitzer, MD and Muls, E 1999, 'Sociodemographic and nutritional determinants of obesity in Belgium'. *International Journal of Obesity,* Vol. 23, no.1, pp. 1S-9S.

Sugimori, H, Yoshida, K, Izuno, T, Miyakawa, M, Suka, M, Sekine, M, Yamagami, T and Kagamimori, S 2004, 'Analysis of factors that influence body mass index from ages 3 to 6 years: A study based on the Toyama cohort study'. *Pediatrics International*, Vol. 46, pp. 302-310.

Sundquist, J and Johansson, S 1998, 'The influence of socio-economic status, ethnicity and lifestyle on body mass index in a longitudinal study'. *International Journal of Epidemiology*, Vol. 27, pp. 57-63.

Taheri, S 2006, 'The link between short sleep duration and obesity: we should recommend more sleep to prevent obesity'. *Archives of Disease in Childhood*, Vol. 91, pp. 881-884.

Visscher, TLS, Kromhout, D and Seidell, JC 2002, 'Long-term and recent time trends in the prevalence of obesity among Dutch men and women'. *International Journal of Obesity*, Vol. 26, pp. 1218-1224.

Vorona, RD, Winn, MP and Babineau, TW 2005, 'Overweight and obese patients in a primary care population report less sleep than patients with a normal body mass index'. *Archives of Internal Medicine*, Vol. 165, pp. 25-30.

Wardle, J and Griffith, J 2001, 'Socio-economic status and weight control practices in British adults'. *Journal of Epidemiology and Community Health*, Vol. 55, pp. 185-190.

World Health Organisation 1998, *Obesity: Preventing and managing the global epidemic. Report of WHO consultation on obesity*. Geneva: World Health Organization.

World Health Organisation 2000, *International association for the study of obesity (IASO) and international obesity task force (IOTF). The Asia-Pacific perspective: Redefining obesity and its treatment.* Geneva: World Health Organization.

World Health Organisation 2011,http://www.who.int/mediacentre/factsheets/fs311/en/index.html Retrieved on April 12, 2012.

World Health Organisation 2016, http://www.who.int/mediacentre/factsheets/fs311/en/ Accessed on October 8, 2017.

Chapter - 8

Growth Pattern and Nutritional Status among the Urban Meitei Girls of Imphal Valley, Manipur

—*R. K. Neeta Devi*

Introduction

The growth of children is considered as one of the most important indicators for understanding the health and well being of a community or nation. Although the paradigm of growth is very stable, its intensity and duration are rather sensitive to environmental agents and the data on growth is a clear indicator that clearly reflects the economy, hygiene and cultural peculiarities (Bodzsar,2000). Various environmental factors such as age, sex, nutrition, socio-economic status, urbanization, etc., influenced growth apart from constitutional hereditary factors (Eveleth and Tanner, 1976; Bogin, 1990; Rao and Busi, 1994).

The growth of children and youth are considered to be best suited for explaining the nutrition and health status of a community (WHO, 1976). Nutrition is the process by which the living organisms receives and utilizes the materials necessary for the maintenance as well renewal of its function. Nutritional status is the physical expression of the relationship between the individual dietary intake, the bioavailability of ingested nutrients and his or her physical requirements (Brown,1984) influenced by the availability of food and purchasing power (Rao et al., 2006).

Malnutrition both underweight and overweight is a direct cause of about half of the more than 10 million child deaths in the developing world each year and approximately 146 million children under five worldwide about one in four are underweight (UNICEF, 2007). Asia and Africa bear the greatest share of all forms of malnutrition of which in 2015, Asia's shares for stunting is 56 percent overweight 48 percent and wasted 68 percent although the progress for stunting within Asia is uneven (Joint Report UNICEF, WHO, and World Bank, 2016). India being a developing country is not only facing the problem of underweight but also the increasing trend of overweight and obesity among children and adolescents due to rising urbanization and improvement in economic development. Overweight and obesity are considered to be a pathological state with multiple causes and is global. However, undernutrition still continues to be the main cause of ill health and premature mortality among children and adolescents (Black et al., 2003; Bose and Bisai 2008a; Nandy et al., 2005; Pelletier, 1994) and holds the higher percentage of nutrition burden in most of the developing nations (Chatterjee and Saha, 2008). It is mainly found in young children in developing countries due to unequal distribution of resource and it mainly follows a vicious cycle that includes poverty and disease which is interlinked in such a way that each contributes to the presence and permanence of the other (Kaur et al., 2005). Undernutrition in children lead to long-lasting physiologic effects including an increased susceptibility to fat accumulation mostly in the central region of the body, lower fat oxidation, lower resting and postprandial energy expenditure, insulin resistance in adulthood, hypertension, dyslipidemia and a reduced capacity for manual work among other impairments and also resulted to poor mental development and school achievement as well as behavioral abnormalities (Martin et al., 2011). Undernutrition during childhood delays growth and its occurrence in the year preceding adolescence delays adolescent spurt (Tanner, 1990) causes lower immune system, thereby increasing the risk of mortality from infectious diseases (Rice et al., 2000).

Anthropometry is one of the best reliable techniques to access the growth and nutritional status of an individual or a population (Gorstein et al., 1994; Jelliffe, 1996).

In Northeast India, few studies have been conducted among different populations on the growth and nutritional status of children and adolescents (Khongsdier, 1996; Khongsdier and Mukherjee, 2003a; Medhi et al., 2006; Medhi et al., 2007; Maken and Varte, 2012; Longkumer, 2013; Mondal and Terangpi, 2014) but these are few in numbers and limited in purpose. There is also dearth of information regarding the growth pattern and nutritional status of urban Meitei girls aged 3 to 9 years of Manipur except a few studies conducted on the nutritional status of Meitei children and adolescents observed the dual occurrence of both underweight and overweight conditions (Bishwalata et al., 2010, Dkhar and Singh, 2012 and Singh and Devi, 2013, Devi and Meitei, 2015; Dkhar and Devi, 2018) but fewer studies are done on the growth pattern of urban Meitei children. Therefore, the present study is an attempt to understand the growth pattern and nutritional status of urban Meitei girls ranging in the age groups of 3 to 9 year.

Materials and Methods

Manipur, a small hilly state of Northeast India, covers a geographical area of 22,327 sq.kms and lies at a latitude of 23°83'N – 25°68'N and longitude of 93°03'E – 94°78'E. Ninety percent of the total geographical area is hilly and inhabited by the Naga and the Kuki Chin tribes and the remaining ten percent is the centrally located oval-shaped valley Imphal Valley inhabited mainly by the Meitei community besides the Muslims (*Meitei Pangals*) and other scheduled castes population. A small percentage of other groups from other parts of the country and the hills have also migrated and settled in the valley.

The Meitei's belong to the Kuki-Chin people of the Mongoloid racial stock and appears to have settled in ancient times in Manipur and the Lushai Hills as well as in the Chittagong Hill Tracts. The written history of the Meities can be traced back to 33 AD. They have their own script known as *Meitei Mayek* and their mother tongue *Manipuri* or *Meiteilon* is the lingua franca of the state. It was included in the 8th schedule by the 71st Amendment of the Constitution in 1992 and it is the only Tibeto-Burman Language in India having a written system.

The staple food is rice and fish is regarded as one of the delicacies of the community. The diet mainly comprises of dried and fermented fish, *ngari* and other fermented food items such as *soibum, soidon, and hawaichar* etc. forms a part of their diet. These are scientifically proven of having high nutritive value and healing power to some ailments. Such type of food habits is also reported in the neighbouring South East Asian countries such as Thailand, Cambodia, Japan, Myanmar, etc., They usually consume raw tender leaves, inflorescence, tender stalks, and other plant parts either chopped into small pieces or taken as a whole along with traditional chutney (*ametpa*) or as an ingredient of *singju*. Apart from these, varieties of seasonally available green leafy vegetables, roots, tubers, pulses, lichens also occupy a part of their main diet. They prefer milk and milk products and seasonal fruits. However, in the last few decades, the valley area of Manipur has undergone tremendous change due to rapid urbanization and socio-economic development. It has resulted in a major transition in terms of economy, occupation, food consumption, physical activity levels, and lifestyle behavior. Children and adolescents are more prefer of consuming energy-rich dense food from fast food restaurants, roadside vendors as well as processed packed food other than the traditional ones.

The present cross-sectional study is conducted on 274 urban Meitei girls in the age group of 3 to 9 years residing in Imphal Municipal area of Lamphel Sub-Division under Imphal-West District of the State. No statistical sampling of household and individuals has been applied for the collection of data. Instead, an attempt has been made to include all children who are willing and co-operative. Anthropometric measurements such as weight and height for each subject are taken following the standard techniques of Weiner and Lourie (1981) and Sen (1994). Body weight is measured with a weighing machine in light minimum clothing and is recorded in terms of kilogram up to the nearest of 0.5 Kg. An anthropometric rod is used for recording the height to the nearest of 0.1mm and is recorded in terms of centimeters. The exact age of each subject is recorded as per the information provided by parents and efforts are made to cross-checked it from date of birth certificates or school registers. For the purpose of the present analysis, the age group is made at the one-year interval i.e age group of 3 years includes children of 2.50 to 3.49 years; age group 4 includes children of 3.50 to 4.49 years and so on (Sen, 1994). The absolute growth or increment per year is calculated by subtracting the mean value of the lower age group from that of the next higher age group.

Nutritional status has been assessed by using three nutritional indices such as body mass Index for age, weight for age and height for age. These indices are expressed in terms of Z-scores following the LMS method as per the reference is given by WHO (2006 and 2007). The Z-score of an individual child at a particular age is calculated as follows

Z= when L0; Z= when L= 0;

Where X is the individual measurement and L (the power in the Box-Cox transformation), M (the median) and S (the generalized coefficient of variation) are the reference values corresponding to a particular age and sex.

The classification for Z scores has been recorded with reference to WHO (2006 and 2007). In case of BMI for age, underweight/thinness has been set as Z-Score < –2 SD; for normal, it ranges from -2 SD scores to 0.99 SD scores; for overweight, it ranges from 1SD to 2 SD scores and for obese, it is set as Z scores > 2 SD scores. For the weight for age and height for age, if Z-scores < -2SD, it is considered as moderate stunted and wasted and moreover, if Z-scores < -3 SD scores, it is considered as severely stunted as well as wasted. The data of the present study has been analyzed by using Ms-Excel and SPSS 17.

Results

Weight, Height, and BMI: The mean weight and height of girls (Table 8.1; Fig.8.1.1 and Fig. 8.1.2) at 3 years is about 12.1 kg and 89.38 cm and gradually becomes heavier and taller with age and attains the maximum weight of 23.74 kg and 125.81 cm at 9 years. During the growth period, the

Table 8.1: Mean and standard deviation of weight and height and BMI of Urban Meitei girls (3-9) years

Age (in years)	*N*	*Weight(kg)*			*Height(cm)*			*BMI (kg/m^2)*	
		Mean	*SD*	*Absolute growth (kg/yr)*	*Mean*	*SD*	*Absolute growth (Cm/yr)*	*Mean*	*SD*
3	32	12.10	1.48	-	89.38	4.51	-	15.12	1.18
4	30	13.87	1.95	1.8	96.41	3.6	7.0	14.88	1.49
5	36	15.36	1.81	1.5	102.68	4.61	6.3	14.54	1.04
6	43	17.49	2.01	2.1	109.29	4.58	6.6	14.65	1.41
7	35	19.20	3.25	1.7	115.20	5.56	5.9	14.39	1.44
8	46	20.83	3.21	1.6	120.20	5.64	5.0	14.37	1.57
9	52	23.74	4.36	2.9	125.81	6.88	5.6	14.90	1.74
(3 to 9) years	274	18.15	4.79	11.6	110.41	13.16	36.4	14.68	1.47

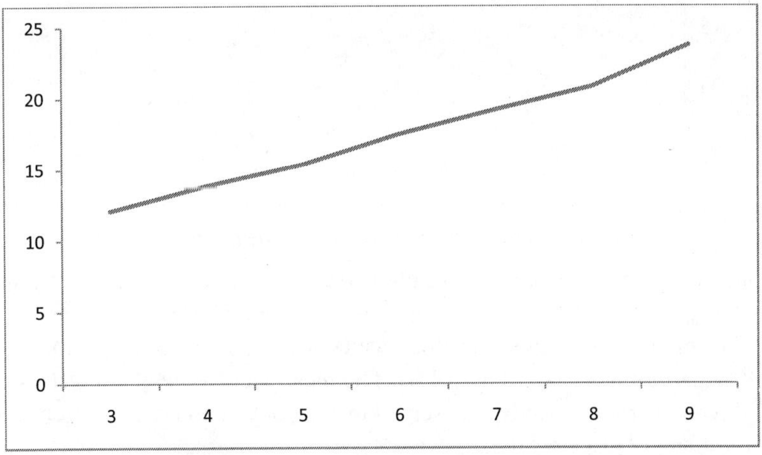

Fig. 8.1.1: Mean distribution of weight (kg) of Meitei girls from 3 to 9 years

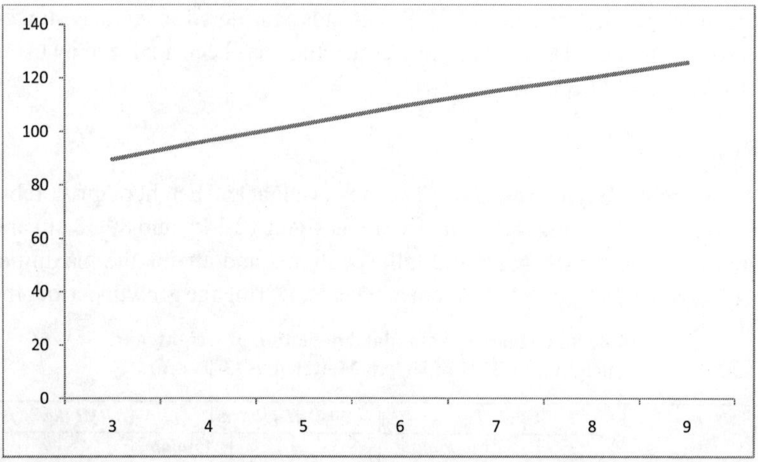

Fig. 8.1.2: Mean distribution of height of Meitei girls from 3 to 9 years

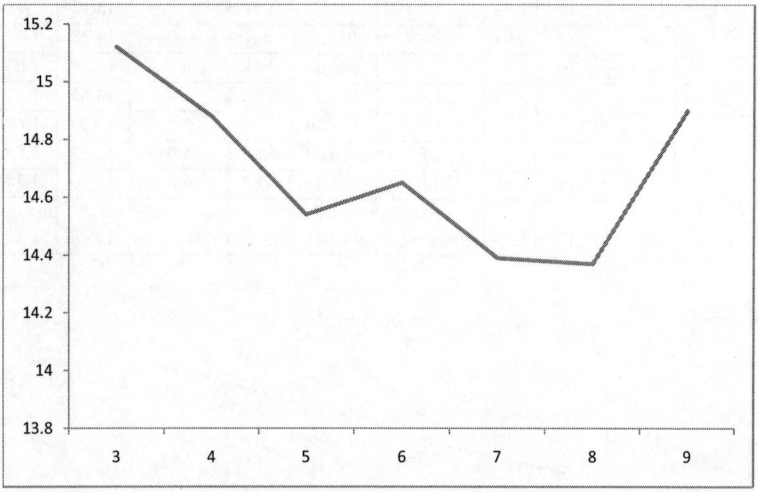

Fig.8.1.3: Mean distribution Body Mass Index of Meitei girls from 3 to 9 years

maximum mean increase in weight is observed in between 8 and 9 years (2.9kg/yr) and for height, it is observed in between 3 and 4 years age groups (7.0 cm/yr). The mean BMI value decelerates with increasing age from 3 to 8 years (Table 8.1 and Fig. 8.1.3). The mean weight, height and BMI of girls in the present study is observed to be about 18.50 kg, 110.41 cm and 14.68 kg/m^2 respectively.

BMI for Age Z scores: The percentage prevalence of underweight and overweight conditions was found to be much lower in all the age groups as compared to the normal condition from 3 to 9 years (Table 8.2; Fig 8.2.1) although slight fluctuations were also noticed in some of the age groups. The prevalence of underweight was 6.6 percent, overweight - 5.1 percent and normal - 88.3 percent (Table 8. 2; Fig. 8.2.2) in the overall age groups, but the distribution of BMI for Age Z scores was found to be insignificant as shown by Chi-square test ($\chi^2 = 10.22$, df=12; P>0.05).

Table 8. 2. Frequency and percentage distribution of BMI for Age Z scores of Metei girls from 3 to 9 years

Age (in years)	*No of subjects*	*Underweight Z scores<-2SD*	*Overweight Z scores = 1to 2 SD*	*Normal Z scores -2SD- <1SD*
		Frequency (%)	*Frequency (%)*	*Frequency (%)*
3	32	1(3.10)	4(12.50)	27(84.40)
4	30	1(3.30)	1(3.30)	28(93.30)
5	36	1(2.80)	0(0.00)	35(97.20)
6	43	3(7.00)	3(7.00)	37 (86.00)
7	35	3(8.60)	1(2.90)	31(88.60)
8	46	5(10.90)	2(4.30)	39(84.80)
9	52	4(7.70)	3(5.80)	45(86.50)
3 to 9 years	274	18 (6.60)	14 (5.10)	242 (88.30)
$\chi^2 = 10.22$, df=12; P>0.05				

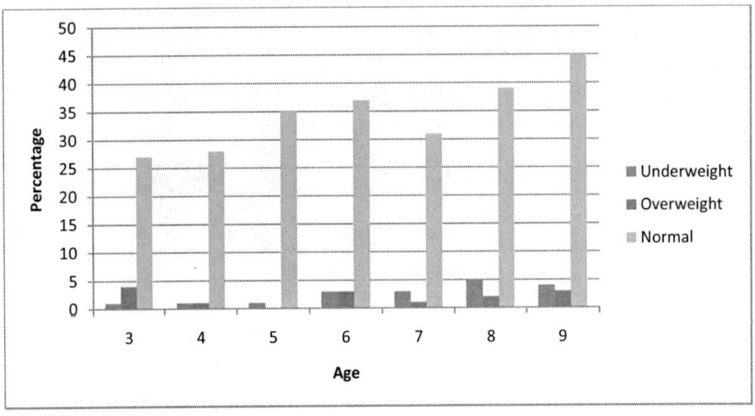

Fig.8.2.1: Distribution of mean BMI Z scores for the age of girls from 3 to 9 years in terms of percentage

Height for Age and Weight for Age Z scores: The percentages of both stunted and wasted conditions were observed to be comparatively much lower as compared to the normal condition in all through the age range of 3 to 9 years (Table 8.3; Fig 8.3.1 and Fig. 8.3.3). Again, it is also observed that in both the stunted and wasted conditions, most of them were moderately

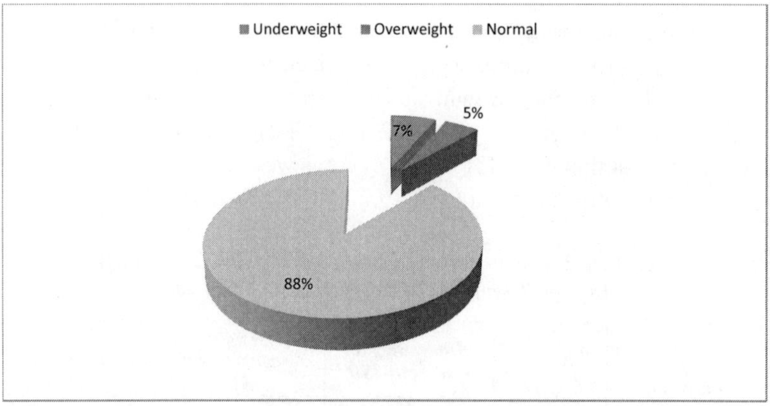

Fig. 8.2.2 Distribution of BMI for age Z scores of Meitei boys in the overall age range from 3 to 9 years

stunted and wasted except a few severe stunted conditions were observed at 6 and 9 years and severely wasted at 8 and 9 years. The prevalence of both stunted and wasted conditions was approximately 15 percent (Table 8.3; Fig 8.3.1) and 18 percent (Table 8.3; and Fig. 8.3.3) respectively. However, Chi-Square test values showed that there was no statistically significant association between age and the prevalence of the stunted condition and wasted conditions.

Table 8.3 Distribution of Height for Age and Weight for Age Z Scores of girls from 3 to 9 years in terms of frequency and percentage

Age (in years)	*No. of subjects*	*Height for Age Z Scores*			*Weight for Age Z scores*		
		Stunted Conditions			*Wasted Conditions*		
		Severe stunted Z scores <-3SD	*Moderate stunted Z scores <-2SD*	*Normal Z scores ≥-2SD*	*Severe Wasted Z scores < -3SD*	*Moderate Wasted Z scores <-2SD*	*Normal Z scores ≥-2SD*
		Frequency (%)	*Frequency (%)*	*Frequency (%)*	*Frequency (%)*	*Frequency (%)*	*Frequency (%)*
3	32	0 (0.00)	10 (31.30)	22 (68.80)	0 (0.00)	5 (15.60)	27 (84.40)
4	30	0 (0.00)	5 (16.70)	25 (83.30)	0 (0.00)	7 (23.30)	23 (76.70)
5	36	0 (0.00)	8 (22.20)	28 (77.80)	0 (0.00)	7 (19.40)	29 (80.60)
6	43	2 (4.70)	2 (4.70)	39 (90.70)	0 (0.00)	2 (4.70)	41 (95.30)
7	35	0 (0.00)	4 (11.40)	31 (88.60)	0 (0.00)	6 (17.10)	29 (82.90)
8	46	0 (0.00)	6 (13.00)	40 (87.00)	1 (2.20)	8 (17.40)	37 (80.40)
9	52	1 (1.90)	7 (13.50)	44 (84.60)	4 (7.70)	8 (15.40)	40 (76.90)
3 to 9 years	274	3 (1.10)	42 (15.30)	229 (83.60)	5 (1.80)	43 (15.70)	226 (82.50)
		χ^2=18.938,df=12; P>0.05			χ^2=19.20,df=12; P>0.05		

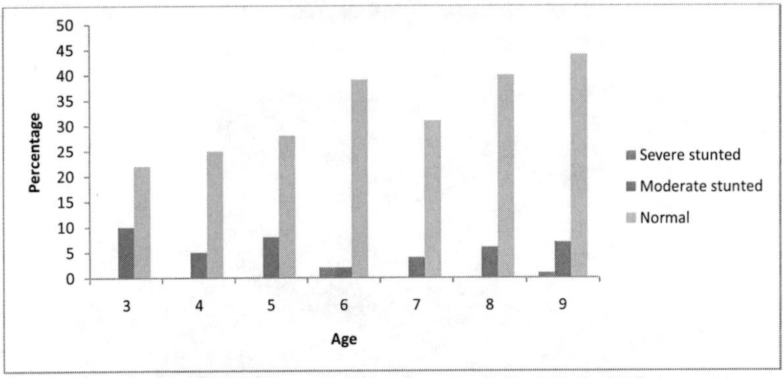

Fig. 8.3.1 Distribution of mean height for age Z scores of girls from 3 to 9 years in terms of percentage

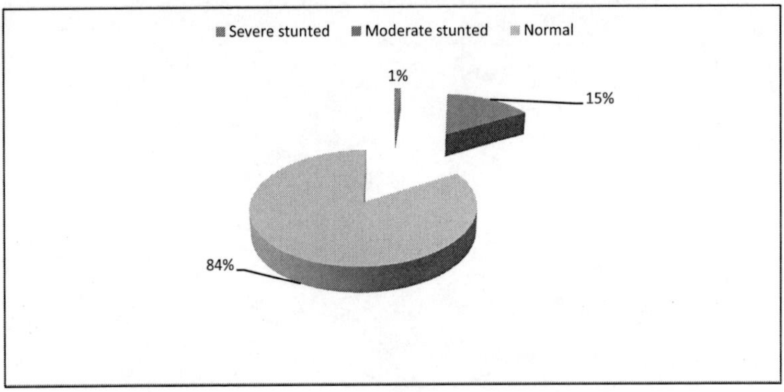

Fig. 8.3.2 Distribution of mean height for age Z scores of girls in the overall age groups of 3 to 9 years in terms of percentage

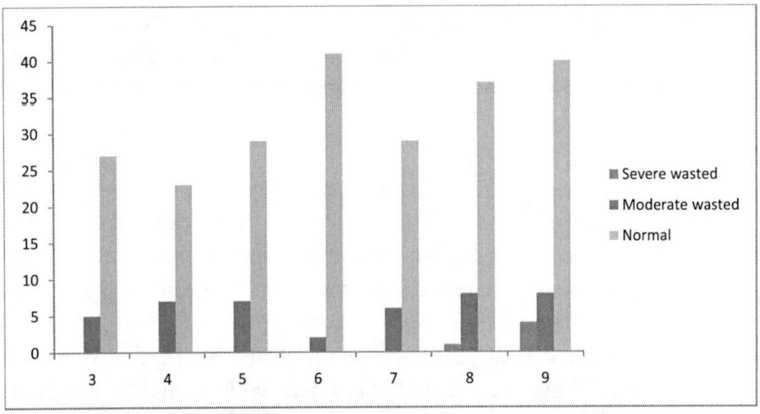

Fig. 8. 3.3 Distribution of weight for age Z scores of Meitei Girls in the age range of 3 to 9 years in terms of percentage

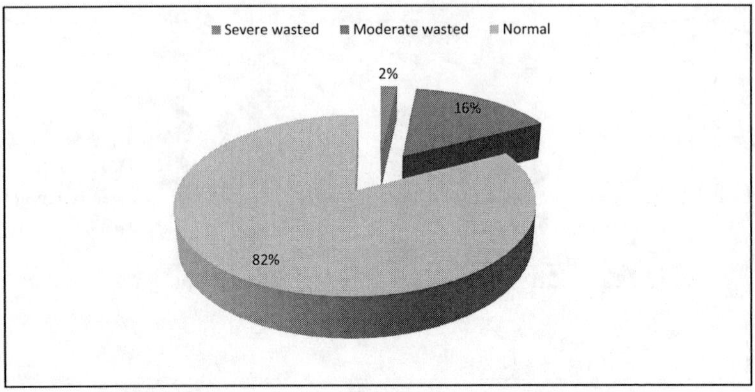

Fig. 8.3.4 Distribution of weight for age Z scores of Meitei girls in the age range of 3 to 9 years in terms of percentage

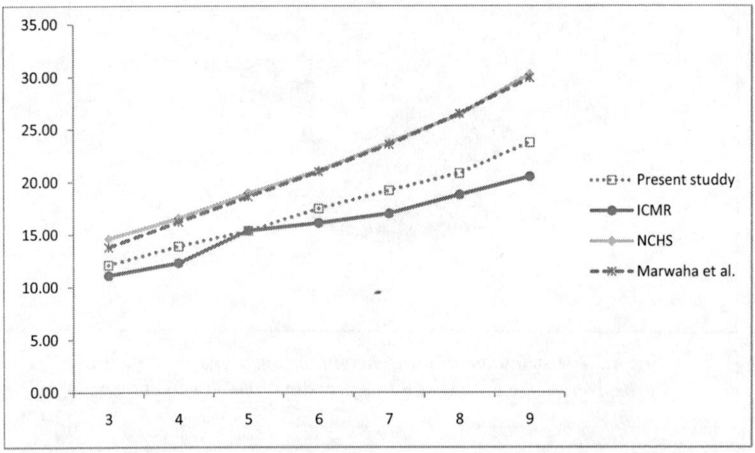

Fig. 8.4.1 Comparisons of the weight of Meitei girls with 50th centile curves of ICMR, NCHS and Indian school girls of high socio-economic strata

Discussion

The weight and height of girls in the present study observed a gradually increasing trend with increasing age from 3 to 9 years. Several cross-sectional studies conducted among different populations of both urban and rural have also found that with the advancement of age, there is a tendency for acceleration of all the physical traits except skin folds (ICMR, 1984; Tanner, 1990; Busi et al., 2003 and Patil et al., 2014). The weight and height of girls in the present study are observed to be lighter and shorter when compared with 50th centile curves of the NCHS standard and Marwaha

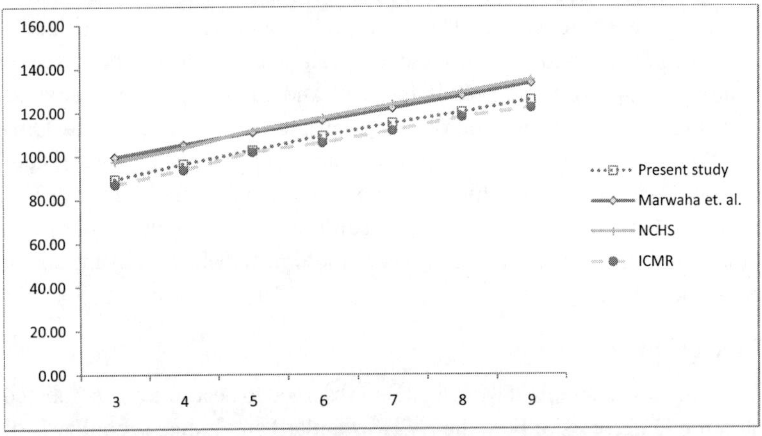

Fig. 8.4.2.Comparison of the height of Meitei girls with 5oth centile curves of ICMR, NCHS and Indian school girls of high socio-economic strata

et al., 2011 curves of Indian school girls of high socio-economic strata but when compared with ICMR pooled urban data, the height was found to be more or less the same but heavier in weight than the former. Growth failures in developing countries like India are mainly due to inadequate nutrition and exposure to infections because of poor socio-economic and hygiene conditions (Tanner, 1989; Floyd, 2000). Even in developed nations, growth is curtailed in some sections of the population by deprivation, illness, psychological stress and family size (Norgan, 2000). It is seen from the findings of the present study that both forms of malnutrition (underweight and overweight condition) coexisted in the population where the underweight percentage (6.6%) is found to be slightly higher than the overweight percentage (5.1%). Developing countries including India are found to be facing nutrition transition where the double burden of both underweight and overweight conditions existed together although underweight condition remains a major contributing factor (Iyer et al., 2011; Jeemon et al., 2009; Singh and Devi, 2013; Unnithan and Syamakumari, 2013). The possible reasons for the higher prevalence of underweight condition might be due to the poverty that contributes to household food insecurity, poor maternal and child care, unhealthy environments and poor access to health services, low dietary intake and chronic infections (Ramzan and Khan, 2008; WHO, 2011). The reasons of overweight in the present study might be due to low levels of physical activity, television viewing, computer games, internet, preference of consuming junk foods, overemphasis on academic excellence and automated transport facilities.

The present study also observed that 16.0 percent of girls are stunted and 18.0 percent are wasted. The causes of growth retardation are deeply rooted in poverty and lack of education (Kaur et al., 2005), poor socio-economic conditions and risk of frequent and early exposure to adverse conditions such as illness and or inappropriate feeding practice (de Onis and Blossner, 1997). Children in low-income groups have a greater risk of stunting than those in the high-income group (Maken and Varte, 2012). Rustein (2000) also opines that the health status of young children and adolescents are influenced by a variety of biological, behavioural and socio-economic variables.

Conclusion

The growth pattern of Meitei girls is observed to be inferior as compared to their counterparts of both the NCHS standard and Indian school girls of upper socio-economic strata (Marwaha et al., 2011) although they are heavier and most of the same height as compared to their ICMR counterparts. The present study population also experienced the coexistence of both forms of underweight and overweight conditions although the former is slightly higher than the later. It can be concluded that the present population is in the early phase of nutrition transition.

References

Banjade, B and Vijaya, AN 2014, 'Prevalence of obesity and its risk factors among pre-university college adolescents of Belgaum City, Karnataka'. *International Organization of Scientific Research, Journal of Dental and Medical Sciences,* Vol. 13, no.4, pp. 56-60.

Bhoite, R and Uma, I 2011, 'The magnitude of malnutrition and iron deficiency anemia among rural school children: An appraisal'. *Asian Journal of Experimental Biological Sciences, Vol.* 2, no.2, pp. 354-361.

Bisai, S, Bose, K, Ghosh, D and De, K 2011, 'Growth pattern and prevalence of underweight and stunting among rural adolescents'. *Journal of Nepal Paediatric Society, Vol.* 31, no.1, pp. 1-24.

Bisai, S, Khongsdier, R, Bose, K, Bhunia, AK, Mahalanobis and, D and Koziel, S 2010, 'Prevalence of overweight and obesity among Bengalee adolescents in Midnapore town, West Bengal, India'. *International Journal of Current Research,* Vol. 10, pp. 74-83. Available from: http://www.journalcra.com, [12 April 2013].

Bishwalata, RK, Singh, AB, Singh, AY, Devi, LU and Singh, RKB 2010, 'Overweight and obesity among school children in Manipur, India'. *The National Medical Journal of India,* Vol. 23, no. 5, pp. 263-267.

Black, RE, Morris, SS and Bryce, J 2003, 'Where and why are 10 Million children dying every year'? *Lancet*, Vol. 361, pp. 2226-2234.

Bodzsar, E B 2000, 'A review of Hungarian studies on growth and physique of children'. *Acta. Biologica Szegediensis*, Vol. 44, no.1-4, pp. 139-153.

Bogin, Barry 1990, *Pattern of human growth.* Cambridge University Press.

Bose, K and Bisai, S 2008a, 'Nutritional status of rural adolescent school children in Paschim Medinipur, West Bengal'. *Indian Pediatrics,* Vol. 6, pp. 515–516.

Brown, N K 1984, 'Measurement of dietary intake'. *Population and Development Review, (Suppl*), Vol. 10, pp. 69-91.

Busi, BR, Rao, BD and Rao, V L N 2003, 'Growth progression in physical and physiological variables among Jalaris of Visakhapatnam District, Andhra Pradesh'. *Anthropologist, Vol.* 5, no. 4, pp. 269-282

Caulfield, LE, de Onis, M, Blössner, M and Black, RE 2004, 'Under-nutrition as an underlying cause of child deaths associated with diarrhoea, pneumonia, malaria, and measles'. *American Journal of Clinical Nutrition, Vol.* 80, pp. 193–198.

Chatterjee, S and Saha, S 2008, 'A study on the knowledge and practice of mothers regarding infant feeding and nutritional status of under-five children attending immunization clinic of a medical college'. *The Internet Journal of Nutrition and Wellness,* Vol. 5, no.1.

Chhatwal, J, Verma, M and Riar, SK 2004, 'Obesity among pre-adolescent and adolescents of a developing country (India)'. *Asia Pacific Journal of Clinical Nutrition, Vol.* 13, no.3, pp. 231-235.

Das, S, and Bose, K 2011, 'Prevalence of thinness using new international cut-off points among the Santhal tribal children and adolescents of Purulia District, West Bengal, India'. *Sri Lanka Journal of Child Health,* Vol. 40, no.3, pp. 105-110. Available from: http://dx.doi.org/10.4038/sljch.v40i3.3507, [3 May 2013].

De Onis, M, and Blossner, M 1997, *WHO database on child growth and malnutrition*, WHO, Geneva.

Dehghan, M, Akhtar-Danesh, N, and Merchant, AT 2005, 'Childhood obesity, prevalence, and prevention'. *Nutrition Journal*, Vol. 4, no.24. Available from: http//: doi:10.1186/1475-2891-4-24, [5,August 2012]

Deshmukh, PR, Gupta, SS, Bharambe, MS, Dongre, AR, Maliye, C, Kaur, S and Garg, BS 2006, 'Nutritional status of adolescents in rural Wardha'. *Indian Journal of Pediatrics*, Vol. 73, no.2, pp. 139-141.

Devi, RKN, and Meitei, LI 2015, Nutritional status of adolescent Meitei girls: An anthropometric assessment. Arunkumar, MC, Krishnakumari, P and Joshila Chanu, T(eds) *Health and Health Culture: N.E. Experience*, 177-187. New Delhi: Akansha Publishing House.

Dkhar, JW, and Singh, MS 2012, 'Prevalence of overweight and obesity among the Meitei adolescent of Manipur, India'. *Anthropologists*, Vol. 14, no. 4, pp. 347-351.

Dkhar, JW, and Devi, RKN 2018, Prevalence of underweight, overweight and stunted conditions among urban Meitei adolescent boys of Imphal-West District, Manipur. Ghosh, S, and Limbu, DK (eds) *Readings in Biological Anthropology,* 195-206. New Delhi: B.R. Publishing Corporation.

Eveleth, P B, and Tanner, J M 1976, *Worldwide variation in human growth*. Cambridge, United Kingdom: Cambridge University Press.

Eveleth, P B, and Tanner, J M 1990, *Worldwide variation in human growth*. Cambridge, United Kingdom: Cambridge University Press.

Floyd, B 2000, 'Can socio-economic factors account for a typical correlation between timing, peak velocity, and intensity of adolescent growth in Taiwanese girls'? *American Journal of Human Biology*, Vol. 12, pp. 102-117.

Gorstein, J, Sullivan, K, Yip, R, de Onis, M, Trowbridge, F and Fajans, P 1994, 'Issues in the assessment of nutritional status using anthropometry'. *Bulletin of World Health Organization*, Vol. 72, pp. 272-283.

ICMR 1984, *Growth and physical development of Indian infants and children. Indian Council of Medical Research Technical Report Series*. No. 18

Iyer, UM, Bhoite, RM and Roy, S 2011, 'An exploratory study on the nutritional status and determinants of malnutrition of urban and rural adolescent children (12-16) years of Vadodara City'. *International Journal of Applied Biology and Pharmaceutical Technology*, Vol. 2, no.1, pp. 102-107.

Jeemon, P, Prabhakaran, D, Mohan, V, Thankappan, KR, Joshi, PP, Ahmed, F, Chaturvedi, V and Reddy, KS 2009, 'Double burden of underweight and overweight among children (10–19 years of age) of employees working in Indian industrial units'. *The National Medical Journal of India*, Vol. 22, no.4, pp. 172-176.

Jelliffe, DB 1996, *Assessment of the nutritional status of the community. Monograph Series No. 53*. WHO, Geneva.

Kaneria, Y, Singh, P and Sharma, DC 2006, 'Prevalence of overweight and obesity in relation to socio-economic conditions in two different groups of school-age children of Udaipur City (Rajasthan)'. *Journal of Indian Academy of Clinical Medicine, Vol.* 7, no.2, pp. 133-135.

Kaur, G, Harmander, SK, Singal, P and Sigh, SP 2005, 'Nutritional status: the anthropometric perspective of pre-school children'. *Anthropologists*, Vol. 7, no.2, pp. 99-103.

Kelishadi, R 2007, 'Childhood overweight, obesity, and the metabolic syndrome in developing countries'. *Epidemiological Review, Vol.* 29, pp. 62–76.

Khongsdier, R 1996, 'Growth Pattern of the War Khasi children of Meghalaya'. *Journal of Human Ecology,* Vol. 7, no.3, pp. 175-180.

Khongsdier, R and Mukherjee, N 2003b, 'Growth and nutritional status of Khasi boys in Northeast India relating to exogamous marriages and socio-economic classes'. *American Journal of Physical Anthropology,* Vol. 122, pp. 162-170.

Kurz, KM 1996, 'Adolescent nutritional status in developing countries'. *Proceedings of Nutrition Society,* Vol. 55, no.1b, pp. 321-331.

Lobstein, T 2004, 'Report to WHO-obesity in children and young people: A crisis in public health'. *International Journal of Obesity,* Vol. 28, pp. 41–43.

Longkumer, T 2013, 'Physical growth and nutritional status among Ao Naga children of Nagaland, Northeast India'. *Journal of Anthropology*, Article ID 291239. 6 pages. Available from: http://dx.doi.org/10.1155/2013/291239 [14 May 2014].

Maken, T and Varte, LR 2012, 'Nutritional status of children as indicated by Z-Scores of the Hmars: A tribe of North-East India'. *Antrocom Online Journal of Anthropology*, Vol. 8, pp. 213-227.

Martins, VJB, Florencio, TMMT, Grillo, LP, Franco, Mdop, Martins, PA, Clemente, APG, Santos, CDL, Vieira, Mdef and Sawaya, AL 2011, 'Long lasting effects of undernutrition'. *International Journal of Environment, Public Health*, Vol. 8, pp. 1817-1846. Available from: www.mdpi.com/journal/ijerph, [30 May 2013]

Marwaha, R K N, Tandon, MA, Ganie, R, Kanwar, SC, Sabharwal, A, Bhadra, K and Narang, A 2011, 'Nationwide reference data for height, weight, and body mass index of Indian school children'. *National Medical Journal of India*, Vol. 24, no. 5, 269–277.

Medhi, GK, Barua, A and Mahanta, J 2006, 'Growth and nutritional status of school-age children (6-14 Years) of Tea Garden worker of Assam'. *Journal of Human Ecology, Vol.* 19, no.2, pp. 83-85.

Medhi, GK, Mahanta, NC, and Mahanta, J 2007, 'Nutritional status of adolescents among tea garden workers'. *Indian Journal of Pediatrics,* Vol. 74, pp. 343-347.

Mondal, N and Terangpi, M 2014, 'Prevalence of undernutrition among tribal adolescents of Karbi Anglong District of Assam'. *Sri Lanka Journal of Child Health*, Vol. 43, no.3, pp. 154-157.

Nandy, S, Irving, M and Gordon, D2005, 'Poverty, child undernutrition, and morbidity: New evidence from India'. *Bulletin of the World Health Organization,* Vol. 83, pp. 210–216.

Norgan, NG 2000, 'Long-term economic and physiological consequences of growth retardation in children and adolescents'. *Proceedings of Nutrition Society,* Vol. 59, pp. 245-256.

Patil, AD, Rokade, SA and Pandhare, SR 2014, 'Height and weight of children from higher socio-economic strata in Western Maharashtra'. *International Journal of Recent Trends in Science and Technology* 1, no. 10, pp. 220–222.

Pelletier, DL 1994, 'The relationship between child anthropometry and mortality in developing countries: Implications for policy, progress, and future research'. *Journal of Nutrition,* Vol. 124, pp. 2047S–2081.

Popkin, BM 2000, 'An overview of nutrition transition and its health implication: The Bellagio Meeting'. *Public Health Nutrition,* Vol. 5, pp. 93-103.

Ramzan, M, Ali, I and Khan, AS 2008, 'Body mass status of school children of Dera Ismail Khan, Pakistan'. *Journal of Ayub Medical College, Abbottabad,* Vol. 20, no. 4, pp. 119-121.

Rao, BD and Busi, BR 1994, 'A cross-sectional study of certain linear, transverse and circumferential dimensions among Chenchu tribal girls of Andhra Pradesh'. *Journal of Human Ecology*, Vol. 5, no. 2, pp. 111-116.

Rao, KM, Kumar, RH, Venkaiah, K and Brahman, GNV 2006, 'Nutritional status of Saharia – A primitive tribe of Rajasthan'. *Journal of Human Ecology,* Vol. 19, no. 2, pp. 117-123.

Rao, KV, Balakrishna, N and Adinarayana, K 1986, 'Critical limits of some anthropometric measurements and indices for the assessment of nutritional status'. *Journal of Nutrition and Dietetics*, Vol. 23, pp. 88-99.

Rice, AL, Sacco, L, and Black, RE 2000, 'Malnutrition as an underlying cause of childhood deaths associated with infectious diseases in developing countries'. *Bulletin of World Health Organization,* Vol. 78, pp. 1207-1221.

Rustein, SO 2000, 'Factors associated with infant and child mortality in developing countries during the 1990s'. *Bulletin of the World Health Organization*, Vol. 78, pp. 1256-1270.

Sen, T 1994, *A Guide to Anthropometry*. Calcutta: The world Press Book Limited.

Singh, J and Mondal, N 2013, 'Assessment of nutritional status: A case of tribal children in Assam, Northeast India'. *Journal of Nepal Paediatric. Society,* Vol. 33, no.1, pp. 1-7.

Singh, MS and Devi, RKN 2013, 'Nutritional status among the urban Meitei children and adolescents of Manipur, North-East India'. *Journal of Anthropology*, Article ID. 983485. Available from: http://dx.doi.org/10.1155/2013/983845, [14 January 2014]

Tanner, JM 1989, *Fetus into man: Physical growth from conception to maturity.* Cambridge MA: Harvard University Press.

Thakre, SB, Mohnae, SP, Ughade, SM, Thakre, SS, Morey, SS and Humne, AY 2011, 'Correlates of overweight and obesity among urban school going children of Nagpur City'. *Journal of Clinical and Diagnostic Research,* Vol. 5, no. 8, pp. 1593-1597.

Tharkar, S and Vijay, V 2009, 'Impact of Socio-economic status on the prevalence of overweight and obesity among children and adolescents in urban India'. *The Open Obesity Journal,* Vol. 1, pp. 9-13.

UNICEF 2007, *Growth monitoring and the promotion of healthy young child growth: Evidence of effectiveness and potential to prevent malnutrition.*

UNICEF/WHO/ World Bank Group 2016, *Joint report on levels and trends in child malnutrition.*

Unnithan, A and Syamakumari, S 2013, 'Prevalence of overweight, obesity and underweight among school going children in rural and urban areas of Thiruvananthapuram educational district, Kerala State (India)'. *The Internet Journal of Nutrition and Wellness,* Vol. 6, no.2, pp. 1-7.

World Health Organisation 1976, *New trends, and approaches in the delivery of maternal and child care in health services. WHO Technical Report Series* 600, Geneva.

World Health Organisation 2006, *The WHO multicentre growth reference study. Geneva*: World Health Organization.

World Health Organisation 2007, *The WHO multicentre growth reference study. Geneva*: World Health Organization.

World Health Organisation 2011, *Childhood nutrition and progress in implementing the international code of marketing of breast milk substitutes*. WHO, Geneva, EB 109/11.

Chapter - 9

A Study of Nutritional Status among the Biate Boys of East Jaintia Hills, Meghalaya

— Johnsly Wells Dkhar and Kenneth Darnei

Introduction

Nutritional status is the condition of health of the individual as influenced by the utilization of nutrients. Undernutrition is a major public health problem worldwide, particularly in developing countries (Onis et al., 2000). One of the major health problems in many developing countries is malnutrition (undernutrition and overnutrition) which creates a lasting effect on the growth, development and physical fitness of a person. Malnutrition is defined as 'a state of impaired functional ability or deficient structural integrity or development brought about by a discrepancy between the supply to the body tissues of essential nutrients and the specific demand for them' (Burton, 1976). Undernutrition during infancy and early childhood receive most emphasis because of its association with morbidity and mortality early in life and its long-term consequences; however, undernutrition can occur at any time during the lifecycle. Greater differences are seen when measures are related to age (height-for-age) than when other indices are used (weight-for-age). This indicated that children from developing countries are proportionately smaller than their peers from either developed nations or upper socio-economic status groups in their own country.

Furthermore, the term overweight and obesity, on the other hand, are often used interchangeably, but they are not synonymous. It has been shown that overweight or obesity assessed by the Body Mass Index (BMI) during childhood is a strong predictor of overweight or obesity and coronary heart disease risk factors in young childhood (Janssen et al., 2005). In recent years developing countries including India are facing the double burden of undernutrition and overnutrition as a result of urbanization and improvement in economic development. Malnutrition in developing countries has been connected to a considerable increase in the risk of morbidity and mortality. Indeed, the penalties of malnutrition are frequently expanded not only into later life but into future generation as well (Lobstein et al., 2004). Anthropometry is established worldwide as a non-invasive and less expensive practical technique for nutritional assessment especially in nutritional assessment and epidemiological investigations (WHO, 1995) and Body Mass Index (BMI) is considered to be the best variable for the anthropometric evaluation in nutritional and the general health screening. On the basis of this index the relative proportion of normal, undernourished and obese people can be assessed (WHO, 1995). Assessment of nutritional status is considered as a measure of health and it is necessary for planners to understand the food and nutrition situation among the tribal population for the upliftment of these vulnerable groups. Therefore, the present study is conducted with a view to assessing the nutritional status according to Body Mass Index (BMI), weight and height among the Biate boys aged 3 to 10 years.

Materials and Methods

Meghalaya is a state in Northeast India. The name means 'the abode of clouds' in Sanskrit. The population of Meghalaya is estimated to be 29, 64,007. Meghalaya covers an area of approximately 22,430 square kilometers, with a length to breadth ratio of about 3:1 and has a literacy rate of 75.48 percent. The state is the wettest region of India, recording an average of 12,000 mm (470 in) of rains a year. About 70.0 percent of the state is forested. There are 11 districts in Meghalaya. The Khasis are the largest group, followed by the Garos then the Jaintias. Other groups include the Koch, the Biate's, the Boro, Hajong, Hmar, Karbi, Rabha, and Nepali.

The Biate's are one of the Mongoloid tribes of the Tibeto Burman linguistic groups known as the Mizo Chinnlung or Sinlung tribe of Northeast India. The Biate is one of the oldest hill tribes living in Northeast India especially among the Chin-Kuki-Mizo family. They are akin to Lushai, Hmars, Hrangkhols, Darlongs, Sakchep, Choral, Bawns and other Chin-Kuki Lushai tribes. The Biates have a distinct culture, language, and customs from other tribes. The Biate tribes are distributed in many parts of Northeast India. However, the main concentrated area is in East Jaintia Hills District of Meghalaya and Dima Hasao District of Assam.

The present cross-sectional study is carried out from the Biate 's villages which fall under Saipung Cand RD Block of East Jaintia Hills, Meghalaya. A sample of 255 male subjects in the age range of 3 to 10 years was randomly selected. All apparently healthy and normal males were measured for height and weight following the standard techniques is given by Weiner and Lourie (1981). Body weight of each subject was taken in minimum clothing with the help of weighing machine. An anthropometric rod in terms of centimeters was used for measuring the height of the subject. Body Mass Index (BMI) has been calculated as BMI = Weight (kg)/Height (in meter^2). The assessment of nutritional status for BMI for age, Weight for age and Height for age of the present study which is expressed in terms of Z scores are calculated using LMS method (WHO 1998).

Results

From the present study it is observed that the mean body weight as well as the height are lightest and shortest at the lowest age group of

Table 9.1: Mean Weight, Height, and BMI by age of the Biate Boys

Age Groups (in years)	*N*	*Weight(kg)*		*Height(cm)*		*BMI(kg/m^2)*	
		Mean	*SD*	*Mean*	*SD*	*Mean*	*SD*
3	31	9.77	1.25	79.37	3.11	15.48	1.54
4	30	11.08	2.05	94.63	4.38	13.07	1.41
5	34	13.05	1.51	101.27	5.22	12.77	1.58
6	31	14.93	1.28	110.30	5.52	12.32	1.28
7	34	18.14	2.59	113.93	6.30	13.96	1.52
8	33	21.60	2.49	118.00	5.54	15.52	1.52
9	30	22.83	2.79	122.30	4.53	15.21	1.19
10	32	26.34	2.50	128.37	2.38	15.97	1.29

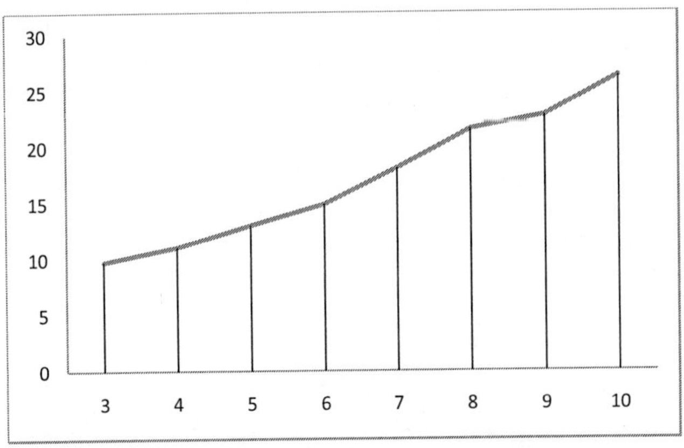

Figure.9.1: Distribution of mean Weight of Biate boys

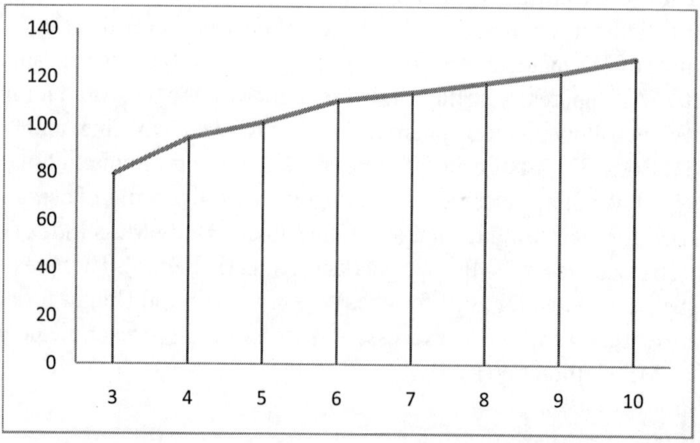

Figure.9.1.1.Distribution of mean Height of Biate boys

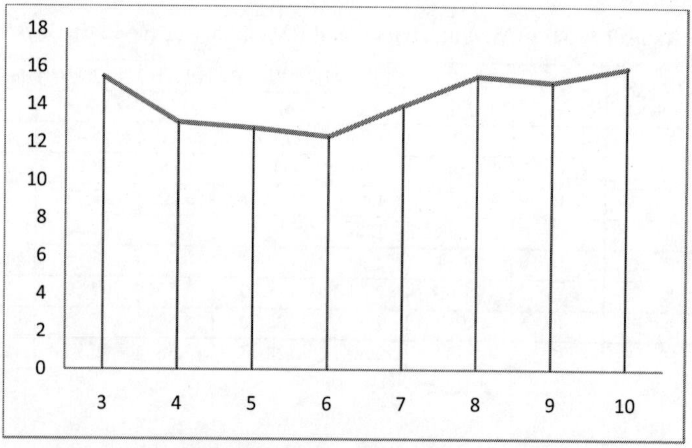

Figure 9.1.2.Distribution of mean BMI of Biate boys

3 years (9.77 kg and 79.37 cm) and heaviest and tallest at the age of 10 years (26.34 kg and 128.37 cm) respectively as shown in Table 9.1 whereas in the case of BMI it fluctuate and can be observed that it is highest at the age of 10 years and lowest at the age group of 6 years.

BMI for age Z-Scores

The distribution of children according to different levels of nutritional status with respect to Body Mass Index (BMI), following the Z-Score of -

2 as a cutoff point is shown in Table 9.1.1 and Figure 9.1.3. It is seen that 64.31 percent are under normal condition, while 10.2 percent are under the moderate condition and 25.49 percent falls under a severe condition. In the overall age groups from 3 to 10 years, it is observed that the normal condition is comparatively higher in percentage as compared to that of the moderate underweight and severely underweight.

Table 9.1.1. Distribution of BMI for age Z-Scores of Biate Boys in the overall age ranges from 3 to 10 years

	Nutritional Status	*Ranges*	*No of Individual*	*Percentage*
	Normal	≥ 2 Z-Score	164	64.31
	Moderate	< - 2 to -3 Z-Score	26	10.20
	Severe	≥ -3 Z-Score	65	25.49
TOTAL			255	100.00

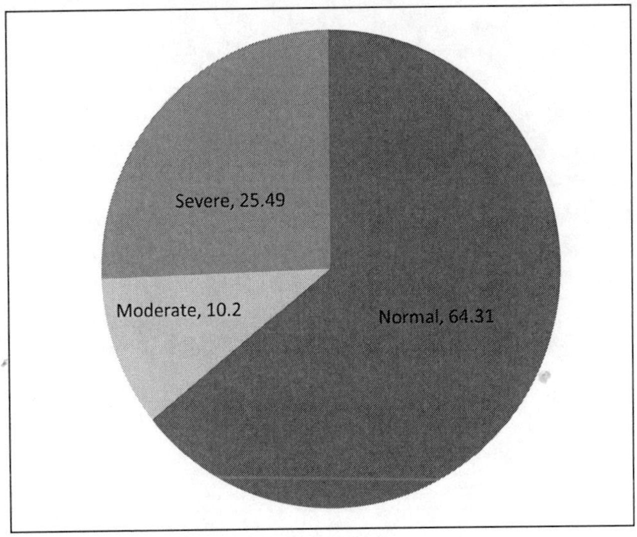

Figure 9.1.3. Distribution of BMI for age Z-Scores of Biate Boys in the overall age range from 3 to 10 years

Weight for age Z-Scores

The nutritional status according to Weight for Age is shown in Table 9.1.2 and Figure 9.1.4. It is seen that 42.35 percent of boys of the present study are gaining their weight according to their age. While about 25.88 percent of boys are under moderate underweight according to their age.

And 31.77 percent are found to be having a severe case which means they are severely underweight according to their age. It is also observed that about 57.65 percent of boys of the present study are suffering from a moderate and severe underweight condition which is higher than the normal weight according to their age.

Table 9.1.2: Distribution of Weight for age Z-Scores of Biate Boys in the overall age range from 3 to 10 years

	Nutritional Status	*Ranges*	*No of Individual*	*Percentage*
	Normal	≥ 2 Z-Score	108	42.35
	Moderate	< - 2 to -3 Z-Score	66	25.88
	Severe	≥ -3 Z-Score	81	31.77
TOTAL			255	100.00

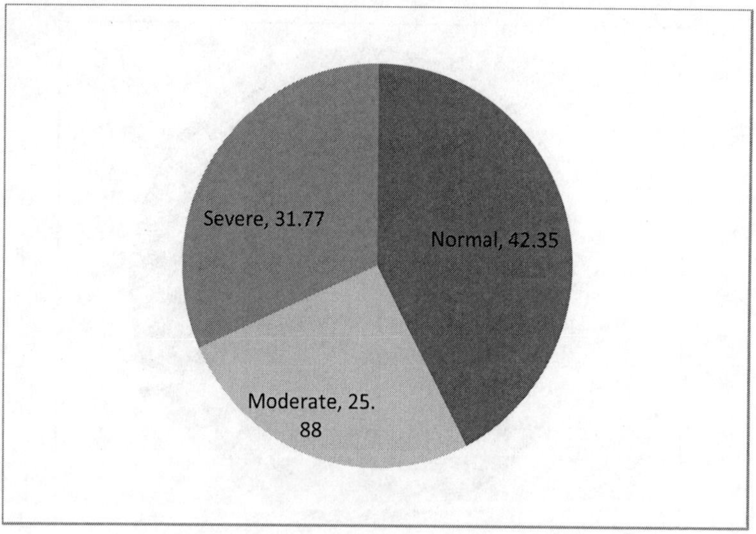

Figure 9.1.4: Distribution of Weight for age Z-Scores of Biate Boys in the overall age range from 3 to 10 years

Height for Age Z-Scores

It is seen from the Table 9.1.3 and Figures 9.1.5 that 57.25 percent of the boys are found to gained their height according to their age and 23.14 percent are found to be in moderate stunted condition while 19.61 percent of boys are found to be are found to be in severe stunted conditions.

Table 9.1.3: Distribution of Height for age Z-Scores of Biate Boys in the overall age range from 3 to 10 years

	Nutritional Status	*Ranges*	*No of Individual*	*Percentage*
	Normal	≥ 2 Z-Score	146	57.25
	Moderate	< - 2 to -3 Z-Score	59	23.14
	Severe	≥ -3 Z-Score	50	19.61
TOTAL			255	100.00

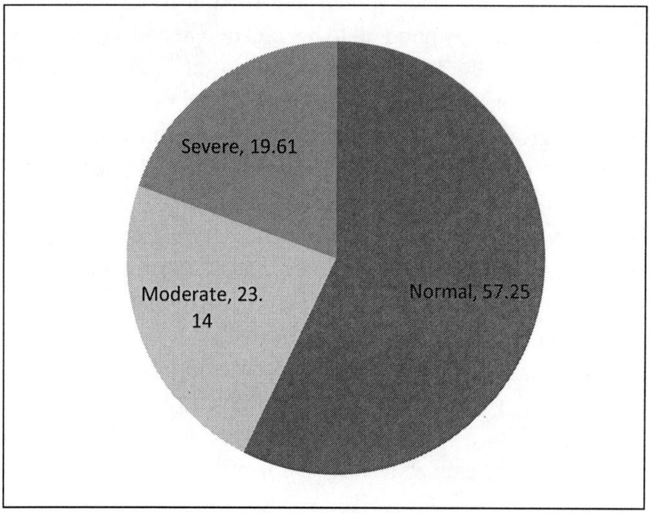

Figure 9.1.5: Distribution of Weight for age Z-Scores of Biate Boys in the overall age range from 3 to 10 years

Discussion

Anthropometric examination is appropriate means in any research to evaluate health and nutritional condition in children and the indices does not only directly reflect the socio-economic status of the family, health and social well-being of the population, but also the competence of the health care system, and the influence of the immediate surroundings, and can also predict academic performance (Srivastava et al., 2012). As far as the present study is concerned, three anthropometric indices namely weight for age, height for age and BMI for age are taken into consideration. As mentioned, weight for age is considered as a measure for underweight and overweight and height for age is considered an indicator of growth retardation and stunting, while on the other hand BMI for age is considered a good indicator of fatness and thinness or wasting due to Chronic Energy Deficiency. From the present study it is observed that the prevalence of growth retardation assesses by weight for age is found to be 57.65 percent and height for age is found to be 42.75 percent and the prevalence of Chronic Energy Deficiency as measured by BMI (Body Mass Index) is found to be 35.67 percent among the Biate boys of East Jaintia Hills Meghalaya . It is also observed that the weight and height increase with the increase in age, whereas in the case of BMI it fluctuate and it is highest at the age of 10 years and lowest at the age group of 6 years. This growth retardation may be due to anomalies in food supplies and food utilization and the growth performance is relatively good up to 3 years of age and again from 7 years of age in respect to the BMI, perhaps due to the reasons that the infant depends mainly on maternal food supplies of breastfeeding and the retardation of growth between 4 to 7 years of age may be related to low socio-economic condition. It is also observed that 42.35 percent of boys of the present study are gaining their weight according to their age and 31.77 percent are found to be having severely underweight according to their age. The present study has also observed that 57.25 percent of the boys are found to gain their height according to their age and 23.14 percent are moderately stunted and 19.61 percent are severely stunted. The present sample has a relatively greater tendency to be on the underweight side in comparison to being on the overweight side. According to Freedman (2000) and Kopelman (2000), lower values of BMI are associated with morbidity and also increase the expectancy of health risks.

Conclusion

The present study highlighted that the nutritional standards of the Biate boys unsatisfactory. The study has also revealed the existence of prevalence of malnutrition and also co-existence of underweight and stunted conditions which shows the need for further studies related to nutritional and health status in order to know the exact cause of the prevalence of malnutrition.

References

Bareh, H 1974, *Meghalaya.* Shillong: Northeastern India news and feature service.

Burton, BT 1976, *Human nutrition*, 3rd ed. Mc Graw-hill, New York.

De Onis, M, and Blössner, M 2000, 'Prevalence and trends of overweight among preschool children in developing countries'. *The American Journal of Clinical Nutrition*, Vol. *72*, no. 4, pp. 1032–1039.

Freedman, JM 2000, 'Obesity in the new millennium'. *Nature,* pp. 632-634.

Janssen, I, Katzmarzyk, PT, Boyce, WF, Vereecken, C, Mulvihill, C, Roberts, C and Pickett, W 2005, 'Comparison of overweight and obesity prevalence in school-aged youth from 34 countries and their relationships with physical activity and dietary patterns'. *Obesity Reviews*, Vol. 6, no.2, pp. 123–132.

Kopelman, PG 2000, 'Obesity as a medical problem'. *Nature,*pp.635-643.

Lobstein, T, Baur, L, and Uauy, R 2004, 'Obesity in children and young people: a crisis in public health'. *Obesity Reviews*, Vol. 5, no.s1, pp. 4–85.

Pakem, B 1980, *The Bates. A seminar paper on tribes in Northeast India,* September 22-24, Organized by Northeast India Council for Social Science Research, Shillong.

Srivastava, A, Mahmood, SE, Srivastava, PM, Shrotriya, VP and Kumar, B 2012, 'Nutritional status of school-age children- A scenario of urban slums in India'. *Archives of Public Health*, Vol. 70, no.1, pp. 8.

Weiner, JS, and Lourie, JA. 1981, *Practical human biology*. London: Academic Press.

World Health Organisation 1995, *Physical status: The use and interpretation of anthropometry. WHO Tech. Rep.* Ser. No. 854. WHO, Geneva.

World Health Organization Working Group on the Growth Reference Protocol 1998, *A growth curve for the 21st century: The WHO multicentre growth reference study* (protocol). Geneva:World health organization.

Chapter - 10

Body Mass Index in Relation to Socio-economic Factors among the Hmars of Northeast India

— Temsutola Maken and L. Robert Varte

Introduction

The Hmars are one of the major tribes of Northeast India and they are distributed in Churachandpur, Tipaimukh and Jiribam divisions of Manipur state. A good number of Hmars are also found in North Cachar Hills (Assam), Meghalaya, Mizoram, and Tripura. The present study deals with the Hmars of Aizawl district in Mizoram. The term 'Hmar' is believed to have originated from the term 'Hmarh' meaning 'tying of one's hair in a knot on the nape of one's head'. Physically, the Hmars are a strong and sturdy race and belong to the Mongoloid stock, with a flat broad nose and short stature, and well-developed legs. The skin is usually yellow-brownish in colour. The face is broad and flat and body is fairly well built with scanty body hairs. The men seldom have hair on their faces (Allen et al.,1979).

A number of monographs describe the socio-cultural life of these hill tribals (Songate, 1957, 1976; Allen et al.,1979). However, with regard to anthropometric data and especially nutritional status very little work has been done so far (Maken and Varte, 2012; Varte et al., 2015; Lalnuneng and Khongsdier, 2017). With this in mind, the present work was undertaken to study the anthropometric characteristics and nutritional status of the Hmars of Aizawl District, Mizoram.

From the biological anthropological point of view, demographic variables (e.g., fertility, mortality and reproductive wastage), self-reported morbidity, nutritional anthropometry of adults, growth and nutritional status of children, and haematological traits including blood pressure and haemoglobin contents are commonly used as health indicators of the population (Khongsdier, 1996). These health variables/traits are in turn influenced by numerous biosocial factors. An attempt to understand the relationship of these health traits with various socio-economic factors like educational level, occupation, income, etc., may be very helpful in understanding the health status of a population.

BMI as a Measure of Assessing Nutritional Status and Obesity

Obesity was first included in the International Classification of Diseases in 1948. (International Classification of Diseases, 1948). It is now increasing at an alarming rate throughout the world. In 1988, it was estimated that there are more than 250 million obese people worldwide (WHO, 1998). In 2016, more than 1.9 billion adults, 18 years and older, were overweight. Of these, over 650 million were obese. In terms of percentage, these translate to 39 percent of adults aged 18 years and above were overweight and 13 percent obese (WHO, 2018).

The National Family Health Survey report shows that 26.6 percent of urban and 14.3 percent rural men (18.9% of Indian men) and 31.3 percent urban and 15.0 percent rural women (20.7% of Indian women) are either overweight or obese (NFHS, 2015-2016).

Quetelet in 1869 was the first person to observe that among adults of normal build but different heights, weight was roughly proportional to height squared. Quetelet Index (Weight in kg. /Height in meters^2) was later renamed Body Mass Index (BMI) by Keys et al.,1972.

In adult individuals, Quetelet or Body Mass Index (BMI) and upper arm muscle area are widely accepted as one of the best indicators of the nutritional status of adult individuals and/or populations (James et al., 1988; Ferro-Luzzi et al., 1992; Shetty and James, 1994). It is suggested that the BMI may be more nutritionally than genetically related (Rolland-Cachera, 1993), despite the fact that there is a wide variation between human populations in weight and height (Eveleth and Tanner, 1990; Majumder et al., 1990).

BMI measurements facilitate the comparison between past and present populations and across countries, especially given that other measurements are not readily available for poor countries. Even in richer countries, like the US, other obesity tests remain too expensive and prevent the use of large samples.

While there are other alternatives to BMI that are more accurate in terms of measuring areas of regional fat, BMI is a good and reliable predictor of total fat, and it is relatively inexpensive to measure (Allison et al., 2002).

At present, even in developing countries, obesity has been considered to be a health problem that cannot be left unattended to. In India, studies on obesity/BMI especially in understanding the relationship between BMI and socio-economic factors, especially among the populations in Northeast India is very limited, so such studies are felt necessary to understand better the overall health status of a population. Thus, the use of BMI as an anthropometric indicator of nutritional status may be more appropriate in a country with diverse ethnic groups like India.

BMI and Socio-economic Factors

Several studies have revealed the association between BMI and socio-economic conditions (Flegal et al., 1988; Sobal and Stunkard, 1989; Leigh et al., 1992; Millar and Stephens. 1993; Bennett, 1995; Sundquist and Johansson, 1998; Stam-Moraga et al., 1998; Molarius et al., 2000; Osler et al., 2000; Langenberg et al., 2003; Zhang and Wang, 2004; Fei Hu et al., 2005). However, the association between BMI and socio-economic conditions is not always simple and the patterns differ substantially across sex and regions.

For example, lower socio-economic status, (SES) defined by education, income, and housing type was associated with overweight/obesity in Chinese Singaporean women (Sabanayagam et al., 2007). Especially in women, a strong inverse association between obesity and socio-economic status (SES), mostly assessed by educational level, has been reported in numerous affluent populations (Sobal and Stunkard, 1989; Bennett, 1995; Wamala et al., 1997; Stam-Moraga et al., 1998; Wardle and Griffith, 2001). In another study among women, overweight was found to be associated with lower household income (Gortmaker et al., 1993; Sarlio-Lähteenkorva and Lahelma, 1999).

This association was less consistent in men, although in about half of the populations in the 1990s an inverse association was observed (Molarius et al., 2000). In men, SES was found not to be associated with overweight/obesity (Sabanayagam et al., 2007). In most of the other populations, no association between education and BMI was found. The results suggested a positive association only in some eastern and central European populations (Molarius et al., 2000).

While in some regions we find a positive relationship between BMI and socio-economic status, in some other region, the opposite is reported.

For example, in a study reported from Brazil, the risk of obesity was reported to be strongly increased with income in both the more developed and less developed regions (Monteiro et al., 2001). This relationship between BMI and socio-economic factors also differ among males and females of the same population (Sanz-de-Galdeano, 2005).

Overall, associations between SES and obesity, and the consistency of this association appear to vary across countries, perhaps on the basis of affluence of the country (Molarius et al., 2000). It varies among men and women too. In women only, this risk was found to be associated with low or primary education levels, being unemployed or working at home and having a medium income (Amelia et al., 2008).

In India, (Bharati, 1989; Rao et al., 1990) has observed that almost all anthropometric measurements and indices tend to be higher in the higher income groups. Similar reports of a direct, positive association between overweight/obesity and income have been made among the south Indian population (Reddy, 1998; Kaneria et al., 2006). In a study in rural India to identify social and economic factors associated with a difference in BMI, it was reported that women were thinner in joint land-owning families, where the main occupation was farming than those in non-farming families. This was not true of men in this type of family (Barker et al., 2006).

In Northeast India, there are only a few studies on the relationship between adult anthropometry/BMI and socio-economic conditions (Khongsdier, 1997, 2002, 2005c). Existing literature shows that BMI is increasing both in developed and developing countries. This increase is also quite interesting in the fact that difference is seen among the sexes. Socio-economic factors play a very important role in regulating obesity because there is a difference in education, income, and occupation that regulates obesity differently in both men and women.

BMI and Educational Level

The relation between education and BMI have been studied by many and though education enables people to integrate healthy behaviours (eg specific dietary pattern, exercise), into a coherent lifestyle, gives them a sense of control over their health (Mirowsky and Ross, 1998) and limits exposure to negative influences associated with the social and physical environment in which one lives and works (Winkleby et al., 1990). This relationship between education and BMI is still not uniformly similar among all populations.

For example, in a Finnish study, subjects with low education were more likely to gain weight than well-educated men and women (Rissanen et al., 1991). In another study conducted in Norway, women with a lower level of education had a significantly greater BMI than more educated women. In both 1990 and 2001, rural men with the highest level of household income had a greater BMI than rural men earning less income. Weight gain occurred across all education and income brackets, with no differential associations between SES strata and changes in BMI for either gender or (Reas et al., 2007).

However, in Sweden, less educated men were shown to gain less weight than men with higher education (Sundquist and Johansson, 1998).

BMI and Income

Income may reflect access to medical care recourses, good housing, and working conditions, and provides opportunities for healthy lifestyles. Also, higher SES is positively associated with weight control behaviours such as physical activity, access to healthy foods, and less time spent watching television. (Sobal and Stunkard, 1989; Jeffrey and French, 1996; Jeffrey et al., 1991; Jeffrey and French, 1998).

In US adults during a 3-decade period (1971-2002), it was reported that obesity has increased at all levels of income. Moreover, it is typically not the poor who have experienced the largest gains. They also reported that increases in the prevalence of obesity are neither limited to nor typically highest among the poor (Chang and Laudrdale, 2005). While attention is often drawn to the prevalence of obesity among the poor, some data suggest that secular changes have occurred among persons with a higher SES as well (Mokdad et al., 1999). In a study carried out in Canada, there was an inverse association between BMI and income. After adjustment for age and sex in a linear regression model, lower BMI was found with the increase in income (Finkelstein, 2008).

Thus, body mass and the prevalence of obesity have been shown to be inversely associated with socio-economic status in the United States and other industrialized countries (Garn, 1986; Sonne-Holm and Sorensen, 1986; Sobol and Stunkard, 1989; Khan et al., 1991; Croft et al., 1992; Gortmaker et al., 1993; Randrianjohany et al., 1993; Hanson and Chen, 2007; Finkelstein, 2008).

This is just opposite in developing countries, where earlier studies investigating the aetiology of obesity in India (Sanjeev et al., 1991; Naidu and Rao, 1994) and other developing countries, such as Brazil (Delpeuch et al., 1994), and Kenya (de Vasconcellos, 1994) shows that BMI or obesity is positively associated with socio-economic status and negatively associated with physical activity level.

In developed countries, where the populations had initially experienced greater consumption of a protein-rich diet followed by a rise in cardiovascular deaths, the more literate upper section of the population are known to eat a more balanced diet, consciously avoiding food items that may contribute to obesity, high cholesterol levels, etc. They also probably exercise more, being more conscious of the need to check their weight.

This has probably been just the opposite in developing countries, including India, where the average population is undernourished. The economically well-off upper layer of the population in developing countries tend to consume relatively more protein and fat-rich items and are also relatively physically much less active, a trend probably typical of the initial stages of industrialization, even in developed countries.

Contrary to popular belief, where obesity is believed to affect mostly those in the higher socio-economic status (SES) group, (Monteiro et al., 2004) reported that this trend is already changing. They wrote 'Obesity in the developing world can no longer be considered solely a disease of groups with higher SES. Secondly, the burden of obesity in each developing country tends to shift towards the groups with lower SES as the country's gross national product (GNP) increases. Third, this shift of obesity towards women with low SES apparently occurs at an earlier stage of economic development than it does for men'.

Thus, keeping in mind this transition in BMI among different socio-economic status, studies as such in India in general, and the Northeast, in particular, is very rare. So, the present study is quite important in this aspect.

Aim: The main purpose of the study is to see how socio-economic factors affect BMI, an indicator of the nutritional status of a population.

Materials and Methods

Study Population

Anthropometric data were collected from 978 adults (males = 494 and females = 484) between the age of 20 to 60+ years old among the Hmars, a tribe residing in Northeast India. The four villages covered are Tinghmun, Zohmun, Sakawrdai, and Ratu in Aizawl District, Mizoram. These four villages (20.0% of the listed villages) were selected according to simple random sampling by using random numbers given by Snedecor and Cochran (1967). According to this sampling method, the list of 20 villages and their population in Aizawl District was first prepared based on the information from the District Gazetteer (TRI, 1992). The random numbers were used for selecting the required number of sample villages to assess the health and nutritional status of adult Hmars, a tribe of Northeast India.

Socio-economic Categories

In the present study, two important socio-economic variables were taken into consideration. These include the monthly income of the households and educational level. These socio-economic variables were classified arbitrarily into different groups and/or categories with a view to understanding their influence on the health traits of the study population. Our classification may be briefly described as follows:

Income Groups: Data on household income were collected directly from the head of the household and they were cross-checked taking into consideration some aspects of socio-economic conditions like housing condition, types of occupation, landholding, and monthly expenditure. The interval estimation based on the standard deviation of the per capita monthly income of the household was adopted for classifying the three economic groups (Khongsdier, 1997), which is as follows:

Above ('X + 4SD/N) = High income group (HIG)
('X – 4SD/N) to (Mean + 4SD/N) = Middle income group (MIG)
Below ('X – 4SD/N) = Low income group (LIG)

Educational Level: The data on educational attainment of individuals in the present study were arbitrarily classified as follows: The category 'No education' includes those individuals who were unable to read and write and those who had no education but could read or write their names. The individuals who attended school up to standard V were grouped into 'Primary level' of education. The individuals with the education of standard VI and above are included in the category of 'Secondary level' of education due to the inadequacy of data.

Anthropometry (Adults)

Some selected anthropometric measurements from the basic list of measurements, recommended by the International Biological Programme (Weiner and Lourie, 1981) were taken into consideration for the purpose of the present study. Anthropometric measurements like height (cm), weight (kg), and sitting height (cm) taken on 978 adults (males = 494 and females = 484) aged 20 years and above of both sexes wearing light apparel.

Besides the above measurements, the following indices were computed for adult males and females in order to assess the nutritional status.

1. Body mass index = weight (kg)/height (m)^2
2. Cormic index or relative-sitting height = sitting height (cm)/ height (cm).

An attempt was made to follow as far as possible the standard techniques of taking the measurements as described in Lourie and Weiner (1981).

Statistical Analysis

The data collected for the present study are quantified and analyzed statistically, using SPSS Window software. The data are presented in terms of means, standard deviation, standard error and proportions or percentages. The differences between two means were tested, using the t-student test, while the differences between more than two means were determined, using one-way analysis of variance (ANOVA). Analysis of covariance was also carried out for testing the differences among means, allowing for the effects of other covariates. The differences between proportions were tested, using the chi-square test. Logistic regression analysis was used for analyzing the effects of age, sex, cormic index education and income on the risk factors of under-nutrition.

Results

Adult Body Dimensions and Health Status

Table 10.1 shows the means and standard deviations of anthropometric traits for adult males by age groups. The mean anthropometric traits are higher significantly in the lower age groups. The same is true for adult females as shown in the same table. Perhaps, this may be due to a secular trend other than chance factors. By secular trend, we mean an increase in anthropometric measurements from one generation to another due to improvement in the economic condition of a given population.

Nevertheless, the point to be noted here is that the main purpose of data presented in Table 10.1 was to understand whether or not age influences anthropometric measurements of adult males and females in the present population. It is evident from the results presented in the tables that individual age should be taken into consideration when carrying out statistical analysis on the effects of socio-economic conditions on anthropometric traits, or assessing the nutritional status of the study population in relation to socio-economic conditions.

Table 10.1: Means and standard deviations of anthropometric traits for adult males and females by age groups

Males age groups in years									
Anthropometric characters	*20-39 yrs (n = 236)*		*40-59 yrs (n = 188)*		*60 + yrs (n = 70)*		*Total (n = 494)*		*F-ratio*
	Mean	*SD*	*Mean*	*SD*	*Mean*	*SD*	*Mean*	*SD*	
Weight (kg)	53.48	6.06	53.70	6.91	47.33	7.38	52.70	6.93	27.12***
Height (cm)	159.70	5.48	160.17	5.16	156.67	5.90	159.45	5.53	11.09***
Sitting height (cm)	84.80	4.25	84.52	4.61	81.95	4.04	84.29	4.45	12.05***
Indices									
Body mass index	20.96	2.01	20.93	2.47	19.24	2.44	20.70	2.33	17.21***
Cormic index	0.53	0.02	0.53	0.03	0.52	0.02	0.530	0.02	3.27*
* p < 0.04, *** p < 0.000									

Females age groups in years									
Anthropometric characters	*20-39 yrs (n = 303)*		*40-59 yrs (n = 139)*		*60 + yrs (n = 42)*		*Total (n = 484)*		*F-ratio*
	Mean	*SD*	*Mean*	*SD*	*Mean*	*SD*	*Mean*	*SD*	
Weight (kg)	47.41	5.69	45.73	7.26	40.86	4.24	46.36	6.35	22.53***
Height (cm)	150.34	5.00	149.31	4.77	146.14	5.40	149.68	5.10	13.71***
Sitting height (cm)	80.22	3.26	79.42	3.27	77.41	4.57	79.74	3.48	13.43***
Indices									
Body mass index	20.96	2.15	20.49	2.89	19.14	1.87	20.67	2.42	11.50***
Cormic index	0.533	0.018	0.532	0.019	0.530	0.032	0.532	0.020	0.80
*** p < 0.0001									

Body Mass Index

Body mass index, calculated as weight in kg divided by height squared in meter, is presented in Table 10.2 according to cormic index or relative sitting height index (sitting height/height) and sex. The subjects are grouped into two groups: those who have a cormic index of <0.53 and those who have a cormic index of >0.53. The subjects are divided according to cormic index categories because it is generally reported that BMI is also influenced by cormic index (Norgan,1994; Khongsdier, 2002). It is observed from the table that the mean BMI is significantly higher in both males and females with the higher cormic index when compared with those having a lower cormic index. This is in contrast to the general observation that BMI is lower in those populations with higher cormic index (Norgan, 1994).

Table 10.2: Mean BMI and standard deviation according to sex and cormic index level

	Cormic index of < 0.53			*Cormic index of ≥ 0.53*			*t-value*
	Number	*Mean*	*SD*	*Number*	*Mean*	*SD*	
Males	177	20.30	2.51	317	20.93	2.20	2.87**
Females	142	20.15	2.39	342	20.67	2.65	2.02*
t-value for sex differences	0.54, p > 0.05			1.35, p > 0.05			

*p < 0.04; **p < 0.004

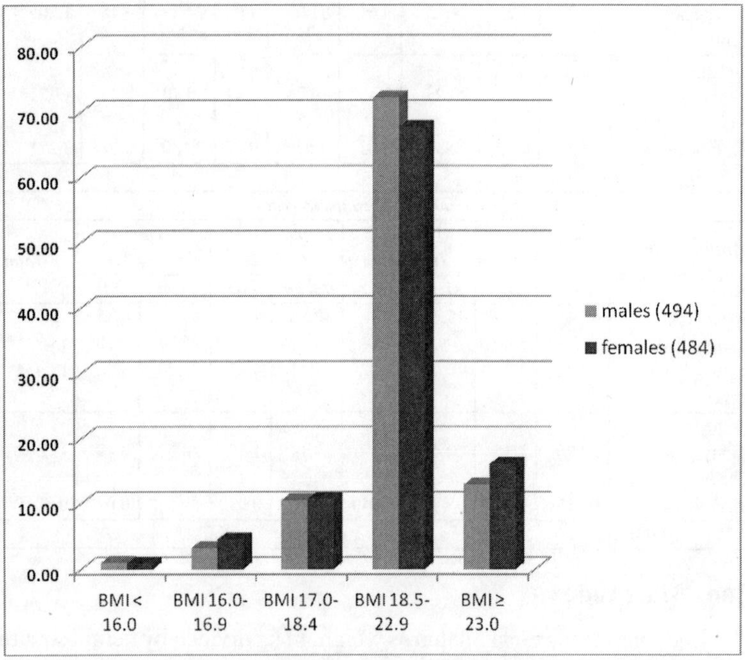

Figure 10.1: Percentage wise distribution of individuals by BMI categories

The percentage distribution of the subjects according to BMI categories is given in Figure 10.1. The individuals with BMI of <16.0, 16.0-16.9, 17.0-18.4, 18.5-23.0 and >23.0 kg/m^2 may be considered as having severe, mild, moderate, normal and overweight grades of nutritional status, respectively (James et al., 1988; WHO, 2000c). It is seen from Figure 10.1 that about 1.01 percent, 3.24 percent and 10.53 percent of adult males suffer from severe, mild and moderate under-nutrition, respectively. Among adult females, these frequencies are 0.83, 4.55 and 10.74 percent, respectively. Considering the cut-off point of 18.5 for screening the individuals into normal and undernourished or chronic energy deficiency (CED) groups (Ferro-Luzzi et al., 1992; WHO, 1995), the overall prevalence of under-nutrition is 14.78 percent in males and 16.12 percent in females. This sex difference in the prevalence of under-nutrition is not statistically significant ($\chi^2 = 0.34$, $p > 0.05$).

With respect to overweight, it has been suggested recently that the BMI of 23.0-25.0 and >25.0 should be considered the cut-off points for screening overweight and obesity, respectively (WHO, 2000c). Therefore, using the cut-off point of 23.0 as a cut-off for screening overweight individuals, Figure 10.1 shows that about 12.96 percent of males and 16.12 percent of females are overweight in the present population. It indicates that the prevalence of overweight is higher in females than in males, although it is not statistically significant ($\chi^2 = 1.97$, $p > 0.05$).

Nutritional Status and Household Income

The relationship between anthropometric traits and socio-economic conditions may be better summarized in terms of anthropometric indices which are frequently used as indicators of nutritional status. As for the adults, BMI is widely used as an indicator of nutritional status. In the present study too, we use BMI in assessing the nutritional status of the adult individuals.

The relationship between mean BMI and economic condition is assessed using per capita income of the household. The results are presented in Table 10.3. It indicates that the unadjusted means of BMI among males (lower income group mean=20.43 is lower than that of the higher income group mean = 21.81). The same is true after adjusting for age, cormic index and education of the mothers (lower income group mean =20.46 is lower than that of the higher income group mean =21.72)

Among the females, the unadjusted means of BMI among females (lower income group mean = 20.32 is lower than that of the higher income group mean = 21.70). The same is true after adjusting for age, cormic index and education of the mothers. (Lower income group mean = 20.33 is lower than that of the higher income group mean = 21.66).

The differences are found to be statistically significant for both the sexes. Thus, it indicates that BMI is positively associated with the income level of the household. The analysis of covariance (ANCOVA) indicates the presence of significant differences especially after adjusting for age, cormic index and income level. In short, the present findings indicate that income is significantly associated with mean BMI.

Table 10.3: BMI according to income groups adjusted for age, cormic index and Education

Income groups	*Unadjusted*			*Adjusted*		
	N	*Mean*	*SD*	*N*	*Mean*	*SD*
Males						
Low Income Group	299	20.43	2.20	299	20.46	2.23
Middle Income Group	102	20.48	1.99	102	20.49	2.22
High Income Group	93	21.81	2.77	93	21.72	2.27
ANCOVA-F ratio	13.53, $p < 0.0001$			11.36, $p < 0.0001$		
Females						
Low Income Group	265	20.32	2.16	265	20.33	2.34
Middle Income Group	123	20.60	2.36	123	20.61	2.34
High Income Group	96	21.70	2.85	96	21.66	2.36
ANCOVA-F ratio	12.09, $p < 0.001$			11.08, $p < 0.001$		

Table 10.4: Prevalence of undernutrition by economic groups

Income groups	*Nutritional status by BMI (kg/m²)*		
	Undernourished (< 18.5)	*Normal*	*Overweight*
Males			
Low Income Group (n =299)	55 (18.39 %)	214 (72.30 %)	30 (10.14 %)
Middle Income Group (n = 102)	10 (9.80 %)	85 (83.33 %)	7 (6.86 %)
High Income Group (n = 93)	8(8.60 %)	58 (60.42 %)	27 (28.13 %)
χ^2 - value	7.93 $p < 0.05$	12.96 $p < 0.001$	25.03 $p < 0.0001$
Females			
Low Income Group (n =265)	50(18.87%)	188 (71.21%)	27 (10.23 %)
Middle Income Group (n =123)	18 (14.63 %)	89 (72.95 %)	16 (13.11 %)
High Income Group (n =96)	11 (11.46%)	58 (59.18 %)	27 (27.55 %)
χ^2 - value	3.17 $p > 0.05$	4.89 $p > 0.05$	17.58 $p < 0.0001$

D.F. = degrees of freedom

With respect to the prevalence of under-nutrition, Table 10.4 shows that about 18.39 percent, 8.30 percent and 8.60 percent of males are undernourished in the low, middle and high-income groups, respectively. These differences are statistically significant in males ($\chi^2 = 7.93, p < 0.05$). In the case of females, the frequencies of under-nutrition are 18.87 percent, 14.63 percent, and 11.46 percent, respectively. Although the prevalence is higher in the lower income groups, the differences are not statistically significant ($\chi^2 = 3.17$, $p > 0.05$).

Nutritional Status and Educational Level

Table 10.5 shows the mean BMI according to educational level. It is observed that mean BMI tends to increase with the increasing educational level. The analysis of covariance (ANCOVA), however, indicates the absence of significant differences especially after adjusting for age, cormic index and income level. In short, the present findings indicate that education is not significantly associated with mean BMI.

Table 10.5: BMI according to educational groups adjusted for age, cormic index and household income

Educational groups	*Unadjusted*			*Adjusted*		
	N	*Mean*	*SD*	*N*	*Mean*	*SD*
Males						
No education	87	20.17	2.56	87	20.51	2.31
Primary	105	20.32	2.27	105	20.54	2.29
Secondary	302	20.99	2.25	302	20.82	2.31
ANCOVA-F ratio	6.06, $p < 0.05$			0.86, $p > 0.05$		
Females						
No education	130	20.48	2.62	130	20.90	2.53
Primary	116	20.68	2.69	116	20.81	2.36
Secondary	238	20.76	2.15	238	20.47	2.51
ANCOVA-F ratio	0.59, $p > 0.05$			1.25, $p > 0.05$		

Table 10.6: Prevalence of undernutrition by educational level

Income groups	*Nutritional status by BMI (kg/m²)*		
	Undernourished (< 18.5)	*Normal (18.5 – 23.0)*	*Overweight (> 23.0)*
Males			
No education (n = 87)	19 (21.84 %)	60 (68.97 %)	8 (9.20 %)
Primary (n = 105)	18 (17.14 %)	74 (70.48 %)	13 (12.38 %)
Secondary and above (n = 302)	36 (11.92 %)	223 (73.84 %)	43 (14.24 %)
Total	73	357	64
χ^2 - value	1.01, $p > 0.05$	1.01, $p > 0.05$	1.56, $p > 0.05$
Females			
No education (n = 130)	29 (22.31 %)	83 (63.85 %)	18 (13.85 %)
Primary (n = 116)	19 (16.38 %)	82 (70.69 %)	15 (12.93 %)
Secondary (n = 238)	31 (13.03 %)	170 (71.43 %)	37 (15.55 %)
Total	79	335	70
χ^2 - value	5.31, $p > 0.05$	2.43, $p > 0.05$	0.48, $p > 0.05$

Table 10.7 shows the prevalence of under-nutrition according to educational level. In general, it indicates that the prevalence of under-nutrition is much higher in those individuals with no educational background as compared to those with primary and secondary levels of education. This holds true for both males and females. It is also observed that the prevalence of overweight is much higher in those subjects with secondary and above the secondary level of education. The prevalence of overweight is about 14.24 percent in males with secondary education as compared to 9.30 percent for those with no education. In females, these frequencies are 15.55 percent and 13.85 percent, respectively. This is according to the theoretical expectation that individuals with higher educational levels are mostly engaged in government and other services, which involve less physical activity, and thereby they are likely to be overweight and obese. However, the chi-square test indicates that the overweight differences between educational groups are not statistically significant (Males: $\chi^2 = 1.56$, $p > 0.05$; Females: $\chi^2 = 0.48$, $p > 0.05$). Thus, from the statistical point of view, it may be difficult to argue that nutritional status is associated with the educational level of adult individuals of the present population.

Table 10.7. Summary of logistic regression analysis for the risk factors of under-nutrition

Parameters	*N*	*Prevalence (%)*	*Odds ratio (95% CI)*	*P-level*
Age groups (years)				
< 40	539	52 (9.65)	-	-
40 - 59	327	60 (18.35)	2.08 (1.37-3.17)	0.001
60 +	112	40 (35.71)	5.06 (2.98-8.61)	0.000
Cormic index				
High (> 0.53)	659	86 (13.05)	-	-
Low (< 0.53)	319	66 (20.69)	1.61 (1.12-2.33)	0.011
Sex				
Male	494	73 (14.78)	-	-
Female	484	79 (16.32)	1.38 (0.95-2.00)	0.089
Income group				
High	194	24 (12.37)	-	-
Middle	224	27 (12.05)	1.04 (0.55-1.97)	0.905
Low	560	101 (18.04)	1.78 (1.04-3.05)	0.037
Educational group				
Secondary	540	67 (12.41)	-	-
Primary	221	37 (16.74)	0.91 (0.57-1.45)	0.806
No education	217	48 (22.12)	1.06 (0.66-1.70)	0.693

In order to have a better understanding of how under-nutrition is related to age, sex, cormic index and socio-economic conditions in the study population, an attempt has been made to carry out the logistic regression analysis as shown in Table 10.7. The odds ratios (with 95% confidence interval (CI) relative to the prevalence of under-nutrition or CED) are derived from logistic regression models after adjusting for other variables under consideration. With respect to age, it is found that older individuals had higher risks of being undernourished after adjusting for a cormic index, sex, education, and income. The individuals aged 60+ years had about 5.06 (95.0% CI: 2.98-8.61) times higher risk of being undernourished as compared with those aged <40 years. On the other hands, the subjects who are aged 60+ years had about 2.08 (95.0% CI: 1.37-3.17) times higher risk in the prevalence of under-nutrition as compared with those aged < 40 years. Thus, it is evident from the present analysis that age is significantly associated with the nutritional status of the individuals in the present population.

As for cormic index, Table 10.7 shows that the individuals with low cormic index seem to have a higher risk of being undernourished. Adjusting for age, sex, education and household income, the prevalence of under-nutrition was about 1.61 (95.0% CI: 1.12-2.33) times higher than those with a cormic index of > 0.53. This indicates that cormic index is significantly associated with the prevalence of CED in the present population. Table 10.7 further shows the odds ratio for the prevalence of CED according to sex. Adjusting for age, cormic index, education, and household income, females had about 1.38 (95.0% CI: 0.95-2.00) times higher risk in the prevalence of CED as compared males. This difference is however not statistically significant ($p > 0.05$).

Table 10.7 further shows the odd ratio for the prevalence of CED according to income groups after allowing for age, sex, cormic index and education. In comparison with the high-income group, the low-income group had about 1.78 (95.0% CI: 1.04-3.05) times higher risk of being undernourished. However, the middle-income group is not significantly deviated from the high-income group with respect to the risk of CED (OR= 1.04; CI: 0.55-1.97). Nevertheless, income seems to have a great influence on the nutritional status of adult individuals in the Hmar population. Like in the case of mean BMI, the effect of education on the nutritional status of the adult males and females is not clearly perceptible in the present population.

Discussion

In the present study, we have taken nutritional status as indicators of the health and nutritional status of Hmar adults. The nutritional status was assessed using body mass index (BMI), which is derived from anthropometric measurements of body weight and height, as internationally recommended (WHO Working Group, 1986; WHO, 1995). These indicators of the health and nutritional status were also correlated with socio-economic variables such as household income and education with a view to understanding the interaction between biological and cultural traits.

Household Income: The relationship between mean BMI and economic condition is assessed using per capita income of the household and it indicates that the unadjusted means of BMI among both the males and females belonging to the lower income group is lower than that of the higher income group. The same is true after adjusting for age, cormic index and education of the mothers.

The differences are found to be statistically significant for both the sexes. Thus, it indicates that BMI is positively associated with the income level of the household. The analysis of covariance (ANCOVA) indicates the presence of significant differences especially after adjusting for age, cormic index and income level. In short, the present findings indicate that income is significantly associated with mean BMI.

Education: From the statistical point of view (Tables 10.6 and 10.7) there is not much difference among different educational levels adjusted for age, cormic index and household income. It may be difficult to argue that nutritional status is associated with the educational level of adult individuals of the present population.

Body Mass Index: The mean values of BMI among the Hmars are higher than those reported for 12 populations in Northeast India (Khongsdier, 2001). It is also higher than those reported for the War Khasi males (Khongsdier, 2002, 2005c). The prevalence of chronic energy deficiency (CED) (<18.5 kg/m^2) was about 14.78 percent in males and 16.12 percent in females. These prevalent rates of CED are also lower than those reported for several populations of Northeast India, including caste groups like Brahmins, Kalitas, Jogis, Kaibartas and Hinduised groups like Ahoms, Kochs, and Rajbanshis (Khongsdier, 2001). They are also lower than the tribal groups like Lalungs, Miris (Khongsdier, 2001) and War Khasis (Khongsdier, 2002).

Prevalence of under- and over-nutrition: The individuals with BMI of <16.0, 16.0-16.9, 17.0-18.4, 18.5-23.0 and >23.0 kg/m^2 were considered to be in the severe, mild, moderate, normal and overweight grades of nutritional status, respectively (James et al., 1988; WHO, 2000c).

In Figure 10.1, it is found that about 1.01 percent, 3.24 percent and 10.53 percent of adult males suffer from severe, mild and moderate under-nutrition, respectively. Among adult females, these frequencies are 0.83, 4.55 percent and 10.74 percent, respectively. Considering the cut-off point of 18.5 for screening the individuals into normal and undernourished or chronic energy deficiency (CED) groups (Ferro-Luzzi et al., 1992), the overall prevalence of under-nutrition is 14.78 percent in males and 16.12 percent in females. This sex difference in the prevalence of under-nutrition is not statistically significant ($\chi^2 = 0.34$, $p > 0.05$).

With respect to overweight, it has been suggested recently that the BMI of 23.0-25.0 and >25.0 should be considered the cut-off points for screening overweight and obesity, respectively (WHO, 2000c). Using the cut-off point of 23.0 for screening overweight individuals, about 12.96 percent of males and 16.12 percent of females are overweight in the present population. It indicates that the prevalence of overweight is higher in females than in males, although it is not statistically significant ($\chi^2 = 1.97$, $p > 0.05$).

The dual presence of under-nutrition and over-nutrition in a population/ state is a complex issue that needs well-planned intervention by policymakers to balance it. As such, the study of both under-nutrition and over-nutrition (obesity) is very important.

Conclusion

Thus, it may be concluded that the nutritional status in the present populations is better than many populations in Northeast India. Considering BMI as an indicator of standard of living in developing countries (Nube, 1998; Khongsdier, 2002), it may be suggested that the Hmar population is also better in economic condition, as indicated by BMI than many other populations in Northeast India. The other possible explanation is that BMI is lower in those populations with higher sitting height (Norgan, 1994; Khongsdier, 2001). However, Hmar adults do not seem to deviate significantly from many populations in Northeast India (Khongsdier, 2001) with respect to cormic index. The higher BMI values in the Hmar population do not seem to be associated with the cormic index, despite the presence of a significant correlation between BMI and cormic index in the present analysis. In other words, although BMI is associated with the cormic index, the Hmars are not much different from other populations in Northeast India with respect to cormic index.

Another point of significance with respect to the findings of the present study is the relationship between BMI and socio-economic conditions. It is observed that mean BMI is positively associated with household income of the household, whereas the prevalence of CED is negatively associated with income. Thus, the suggestion that BMI is an indicator of standard of living seems to have implication for the Hmar population as mentioned above. However, the present study failed to get any significant relationship between BMI and educational level. This indicates that the important factor for BMI is an economic condition of the household, rather than social factors like education.

Acknowledgement

Authors wish to thank all the participants of the four Hmar villages of Aizawl District for their wholehearted support in every possible way throughout the duration of the whole study. Authors also wish to extend their thanks to Prof. R. Khongsdier, NEHU, Shillong for giving immeasurable inputs during the process of writing this paper.

References

Allen, BC, Gait, EA, Allen, CGH and Howard, HF 1979, *Gazetteer of Bengal and Northeast India.* New Delhi: Mittal Publications.

Allison, DB, Zhu, SK, Plankey, M, Faith, MS and Heo, M2002, 'Differential association of body mass index and adiposity with all-cause mortality among men in the first and second national health and nutrition examination surveys (NHANES I and NHANES II) follow-up studies'. *International Journal of Obesity,* Vol. 26, pp. 410-416.

Barker, M, Chorghade, G, Crozier, S, Leary, S and Fall, C 2006, 'Gender differences in body mass index in rural India are determined by socio-economic factors and lifestyle'. *Journal of Nutrition,* Vol. 136, pp. 3062-3068.

Bennett, S 1995, 'Cardiovascular risk factors in Australia: trends in socio-economic inequalities'. *Journal of Epidemiology Community Health,* Vol. 49, pp. 363-72.

Bharati, P 1989, 'Variation in adult body dimensions in relation to economic condition among the Mahishyas of Howrah district, West Bengal, India'. *Annals of Human Biology,* Vol. 16, pp. 529-541.

Chang, VW and Lauderdale, DS 2005, 'Income disparities in body mass index and obesity in the United States, 1971-2002'. *Archives of Internal Medicine,* Vol. 165, pp. 2122-2128.

Croft, JB, Strogatz, DS, James, SA, Keenan, NL, Ammerman, AS, Malarcher, AM, and Haines, PS 1992, 'Socio-economic and behavioral correlates of body mass index in black adults: The Pitt County study'. *American Journal of Public Health,* Vol. 82, no.6, pp. 821-826.

de Vasconcellos, MTL 1994, 'Body mass index: Its relationship with food consumption and socio-economic variables in Brazil'. *Eurasian Journal of Clinical Nutrition,* Vol. 48, no.3, pp. 115-123.

Delpeuch, F, Cornu, A and Massamba, JP 1994, 'Is body mass index sensitively related to socio-economic status and economic adjustment? A case from the Congo'. *Eurasian Journal of Clinical Nutrition,* Vol. 48 (suppl. 3), pp. 141-147.

Eveleth, PB, and Tanner, JM 1990, *Worldwide variation in human growth*, 2nd edition. Cambridge: Cambridge University Press.

Ferro-Luzzi, A, Sette, S, Franklin, M and James, WPT 1992, 'A simplified approach to assessing adult chronic energy deficiency'. *Eurasian Journal of Clinical Nutrition,* Vol. 46, pp. 173-186.

Finkelstein, M 2008, 'The prevalence of diabetes among overweight and obese individuals is higher in poorer than in richer neighbourhoods'. *Canadian Journal of Diabetes,* Vol. 32, no.3, pp. 190-197.

Flegal, KM, Harlan, WR and Landis, JR 1988, 'Secular trends in body mass index and skinfold thickness with socio-economic factors in young adult men'. *American Journal of Clinical Nutrition*, Vol. 48, pp. 544-551.

Garn, SM 1986, 'Family-line and socio-economic factors in fatness and obesity'. *Nutrition Review*, Vol. 44, pp. 381-386.

Gortmaker, SL, Must, A, Perrin, JM, Sobol, AM and Dietz, WH 1993, 'Social and economic consequences of overweight in adolescence and young adulthood'. *The New England Journal of Medicine,* Vol. 329, pp. 1008-1012.

Hanson, MD and Chen, E 2007, 'Socio-economic status, race, and body mass index: The mediating role of physical activity and sedentary behaviors during adolescence'. *Journal of Pediatrics Psychology,* Vol. 32, no.3, pp. 250-259.

International Classification of Diseases, Revision 6, 1948, From (http://www.wolfbane.com/icd/icd6.htm (Accessed November 6, 2015).

James, WPT, Ferro-Luzzi, A and Waterlow, JC 1988, 'Definition of chronic energy deficiency in adults'. *Eurasian Journal of Clinical Nutrition,* Vol. 42, pp. 969-981.

Jeffrey, RW, and French, SA 1998, 'Epidemic obesity in the United States. Are fast foods and television contributing'? *American Journal Public Health,* Vol. 88, pp. 277-280.

Jeffrey, RW, French, SA, Forster, JL and Spry, VM 1991, 'Socio-economic status differences in health behaviors related to obesity: the Health Worker Policy'. *International Journal of Obesity,* Vol. 15, pp. 689-696.

Kaneria, Y, Singh, P and Sharma, DC 2006, 'Prevalence of overweight and obesity in relation to socio-economic conditions in two different groups of school-age children of Udaipur city (Rajasthan)' *Journal, Indian Academy of Clinical Medicine,* Vol. 7, no.2, pp. 133-135.

Keys, AK, Fidanza, F, Karvonen, MJ, Kimura, N and Taylor, HL 1972, 'Indices of relative weight and obesity'. *International Journal of Chronic Diseases,* Vol. 25, pp. 329-343.

Khan, HS, Williamson, DF and Stevens, JA 1991, 'Race and weight change in US women: The roles of socio-economic and marital status'. *American Journal Public Health,* Vol. 81, pp. 319-323.

Khongsdier, R, Nsarangbe, A and Phimu, A 2001, 'A bio-anthropological study of the Pnar of Jatinga in North Cachar Hills of Assam'. *Journal of Indian Anthropological Society,* Vol. 36, pp. 213-222.

Khongsdier, R 1996, 'Assessment of growth and nutritional status: An anthropological perspective'. *Acta Medica Auxologica,* Vol. 28, pp. 147-153.

Khongsdier, R 1997, 'The War Khasi of Meghalaya: Implications of variation in adult body dimensions'. *Journal of Human Ecology, (Spl. Issue)*, Vol. 6, pp. 299-305.

Khongsdier, R 2001, 'Body mass index of adult males in 12 populations of Northeast India'. *Annals of Human Biology,* Vol. 28, no. 4, pp. 374-383.

Khongsdier, R 2002, 'Body mass index and morbidity in adult males of the War Khasi in Northeast India'. *Eurasian Journal of Clinical Nutrition,* Vol. 56, pp. 484-489.

Khongsdier, R 2005c, *Demographic genetics of an Indian population*. Itanagar and New Delhi: Himalayan Publishers.

Lalnuneng, A and Khongsdier, R 2017, 'Rural-urban differences in hypertension among the Hmars of Manipur in Northeast India'. *The Anthropologist*, Vol. 28, no.3, pp. 173-183.

Langenberg, C, Hardy, R, Kuh, D, Brunner, E and Wadsworth, M 2003, 'Central and total obesity in middle-aged men and women in relation to lifetime socio-economic status: evidence from a national birth cohort'. *Journal of Epidemiology Community Health,* Vol. 57, pp. 816-822.

Leigh, JP, Fries, JF and Hubert, HB 1992, 'Gender and race differences in the correlation between body mass and education in the 1971-1975 NHANES I'. *Journal of Epidemiology Community Health,* Vol. 46, pp. 191-196.

Majumder, PP, Shankar, BU, Basu, A, Malhotra, KC, Gupta, R, Mukhopadhyay, B, Vijayakumar, M and Roy, SK 1990, 'Anthropometric variation in India: A statistical appraisal'. *Current Anthropology,,* Vol. 31, pp. 94-103.

Maken, T and Varte, LR 2012, 'Nutritional status of children as indicated by z-scores of the Hmars: A tribe of N.E. India'. *Antrocom Online Journal of Anthropology,* Vol. 8. no. 1. pp. 213-227. ISSN 1973-2880.

Martín, AR, Nieto, JMM, Ruiz, JPN and Jiménez, LE 2008, 'Overweight and obesity: The role of education, employment, and income in Spanish adults'. *Appetite,* Vol. 51, no.2, pp. 266-272.

Millar, WJ and Stephens, T 1993, 'Social status and health risks in Canadian adults: 1985 and 1991'. *Health Science Report,* Vol. 5, pp. 143-156.

Mirowsky, J and Ross, CE 1998, 'Education, personal control, lifestyle, and health. A human capital hypothesis'. *Research on Aging,* Vol. 20, pp. 415-449.

Mokdad, AH, Serdula, MK, Dietz, WH, Bowman, BA, Marks, JS and Koplan, JP 1999, 'The spread of the obesity epidemic in the United States, 1991-1998'. *Journal of American Medical Association,* Vol. 282, pp. 1519-1522.

Molarius, A, Seidell, JC, Sans, S, Tuomilehto, J and Kuulasmaa, K 2000, 'Educational level, relative body weight, and changes in their association over 10 years: an international perspective from the WHO MONICA Project'. *American Journal of Public Health*, Vol. 90, pp. 1260-1268.

Monteiro, CA, Conde, WL and Popkin, BM2001, 'Independent effects of income and education on the risk of obesity in the Brazilian adult population'. *Journal of Nutrition,* Vol. 131, pp. 881S-886S.

Monteiro, CA, Moura, EC, Conde, WL and Popkin, BM 2004, 'Socio-economic status and obesity in adult populations of developing countries: a review'. *Bulletin of World Health Organisation,* Vol. 82, no.12.

Naidu, AN and Rao, NP 1994, 'Body mass index: a measure of the nutritional situation in Indian populations'. *Eurasian Journal of Clinical Nutrition, (Suppl. 3),* Vol. 48, pp. 131-140.

National Family Health Survey (NFHS-4), 2015-16, *National Fact Sheet. INDIA (Provisional data). International Institute for population sciences, Mumbai.* Ministry of Health and Family Welfare. The government of India.

Norgan, NG 1994, 'Population differences in body composition in relation to the body mass index'. *Eurasian Journal of Clinical Nutrition, (Suppl.)*, Vol. 48, pp. 10-27.

Nube, M, Asenso-Okyere, WK and van den Boom, GJM 1998, 'Body mass index as an indicator of standard of living in developing countries'. *Eurasian Journal of Clinical Nutrition,* Vol. 52, pp. 136-144.

Osler, M, Gerdes, LU, Davidsen, M, Bronnum-Hansen, H, Madsen, M and Jorgensen, T 2000, 'Socio-economic status and trends in risk factors for cardiovascular diseases in the Danish MONICA population, 1982-1992'. *Journal of Epidemiology Community Health*, Vol. 54, pp. 108-113.

Randrianjohany, A, Balkau, B and Cubeau, J 1993, 'The relationship between behavioral pattern, overall, and central adiposity in a population of healthy French men'. *International Journal of Obesity,* Vol. 17, pp. 651-655.

Rao, KV, Balakrishna, N, Thimmayama, BVS and Rao, P 1990, 'Indices and critical limits of malnutrition for use among adults'. *Man in India,* Vol. 70, pp. 351-367.

Reas, DL, Nygard, JF, Svensson, E, Sorensen, T and Sandanger, I. 2007, 'Changes in body mass index by age, gender and socio-economic status among a cohort of Norwegian men and women. (1900 -2001)'. *BioMed Central Public Health,* Vol. 30, no.7, 269.

Reddy, BR 1998, 'Body mass index and its association with socio-economic and behavioural variables among socio-economically heterogeneous populations of Andhra Pradesh, India'. *Human Biology,* Vol. 70, pp. 901-917.

Rissanen, AM, Heliövaara, M, Knekt, P, Reunanen, A and Aromaa, A 1991, 'Determinants of weight gain and overweight in adult Finns'. *Eurasian Journal of Clinical Nutrition,* Vol. 45, pp. 419-430.

Rolland-Cachera, MF 1993, 'Body composition during adolescence: Methods, limitations, and determinants'. *Hormone Research, (Suppl.3*), Vol. 39,pp.25-40.

Sabanayagam, C, Shankar, A, Wong, TY, Saw, SM and Foster, PJ 2007, 'Socio-economic status and overweight/obesity in an adult Chinese population in Singapore'. *Journal of Epidemiology,* Vol. 17, no.5, pp. 161-168.

Sanjeev, GD, Indrajit, I and Johnston, FE 1991, 'Skinfold thickness, body circumferences, and their relation to age, sex, and socio-economic status in adults from northwest India'. *Americal Journal of Human Biology,* Vol. 3, pp. 469-477.

Sanz-de-Galdeano, A 2005, *The obesity epidemic in Europe*. IZA Discussion Paper No. 1814.

Sarlio-Lahteenkorva, S and Lahelma, E 1999, 'The association of body mass index with a social and economic disadvantage in women and men'. *International Journal of Epidemiology,* Vol. 28, pp. 445-449.

Shetty, PS, and James, WPT 1994, *Body mass index: A measure of chronic energy deficiency in Adults*. Rome: FAO Food and Nutrition Paper 56.

Snedecor, GW and Cochran, WG 1967, *Statistical methods*, 6th edition. Iowa: The Iowa State University Press.

Sobal, J and Stunkard, AJ 1989, 'Socio-economic status and obesity: A review of the literature'. *Psychology Bulletin,* Vol. 105, pp. 260-275.

Songate, H 1956, *Hmar history-Hmar chanchin*. Imphal: Mao Press.

Songate, H 1967, *Hmar chanchin (Hmar history).*Churachandpur: L and R Press.

Sonne-Holm, S and Sorensen, TIA 1986, 'Prospective study of attainment of the social class of severely obese subjects in relation to parental social class, intelligence, and education'. *The British Medical Journal,* Vol. 292, pp. 586-589.

Stam-Moraga, MC, Kolanowski, J, Dramaix, M, De Henauw, S, De Bacquer, D, De Backer, G and Kornitzer, MD 1998, 'Trends in the prevalence of obesity among Belgian men at work, 1977-1992'. *International Journal of Obesity and Related Metabolic Disorders,* Vol. 22, pp. 988-992.

Sundquist, J and Johansson, SE 1998, 'The influence of socio-economic status, ethnicity and lifestyle on body mass index in a longitudinal study'. *International Journal of Epidemiol*ogy, Vol. 27, pp. 57-63.

Tribal Research Institute 1992, *Gazetteer of villages in Aizawl district*. Aizawl: TRI.

Varte, LR, Sahani, R and Varte, RTh 2015, 'A comparative study of anthropometry and obesity of Hmars of Mizoram and Shillong and blood pressure level among the Hmars of Shillong'. *Journal of the Anthropological Survey of India*, Vol. 63: No. 2, pp. 2333-2344.

Wamala, SP, Wolk, A and Orth-Gomer, K 1997, 'Determinants of obesity in relation to socio-economic status among middle-aged Swedish women'. *Preventive Medicine,* Vol. 26, pp. 734-744.

Xu, F, Yin, XM, Zhang, M, Leslie, E, Ware, R and Owen, N 2005, 'Family average income and body mass index above the healthy weight range among urban and rural residents in regional Mainland China'. *Public Health Nutrition,* Vol. 8, pp. 47-51.

Chapter - 11

Growth, Nutrition, and Physical Activity: An Anthropological Study among the Ao Naga Children from Nagaland, Northeast India

— *Temsumongla Longkumer*

Introduction

Growth refers mainly to the changes in the magnitude of an individual, and the children's physical growth is an important indicator of the health and nutritional status of the human population. Nutrition and physical activity are two important environmental factors in maintaining health and well-being of an individual, and the interaction between them can positively impact on the growth and function of the body as well as positively attribute to one's metabolic fitness. Childhood and adolescence are considered to be a dynamic stage for the assessment of nutrition, as it is a period of growth and development where they undergo physical, mental, emotional, and social changes.

Malnutrition, impairment of health either from a deficiency or excess or imbalance of nutrients, continues to be one of the major health problems, which is of public health significance all over the world. It creates a lasting effect on the growth, development and physical fitness of a person. It is significantly associated with morbidity and mortality early in life and has long-term consequences, although it can occur at any time during the individual's life cycle (Malina et al., 2004a).

Physical activity is a behavior or more appropriately a series of behaviors, which involve bodily movements produced by skeletal muscles, and it comprises a range of behaviors that can be viewed from several perspectives. It is a behavior that occurs within the context of the specific culture within which children and adolescents are reared, and has important biological implications. Numerous factors such as biological, psychological, social, and physical environment may influence activity behavior in children and adolescents (Malina et al., 2004b; Malina and Little, 2008). It is important in maintaining overall health and disease prevention that naturally contributes to physical, mental, and social health, improving the quality of life of people of all ages. Physical inactivity, on the other hand, is indicated as a major risk factor for morbidity and mortality in both children and adults.

In developing countries like India, where there is a drastic change in lifestyle due to the influence of westernization, globalization, and urbanization, the proportion of overweight and obese people now co-exist with those who are undernourished. According to World Health Organization estimates, by the year 2020, non-communicable diseases will account for approximately three-quarters of all deaths in the developing world (WHO, 1997). Nagaland is also experiencing a significant change in many aspects, more specifically in lifestyle including food transition from traditional to modern style with the introduction of fast food joints, restaurants, etc., and research has proved that non-communicable diseases have increased profoundly in the last two decades. Under this backdrop, the present study has been carried out to understand the growth and nutritional status and its association with physical activity among the Ao Naga children.

Materials and Method

A cross-sectional investigation of 790 children aged 8 to 18 years was carried out for the present study (404 boys and 386 girls) from the urban area of Mokokchung district in Nagaland. Seven wards were randomly selected using the lottery method as mentioned in Snedecor and Cochran (1967). No statistical sampling of household and individual were applied for the collection of data. Instead, an attempt was made to include in the sample, all those children within the stipulated age group, who were willing to co-operate.

All the measurements were taken in conformity with the standard definitions and technique laid down by Martin and Saller (1957). Data on height and body weight were collected using anthropometry and weighing machine. Height was measured with an anthropometric rod with a precision of 1mm and weight was recorded to the nearest 0.1 kilograms by weighing machine.

Body Mass Index (BMI) was used to assess nutritional status in the present study. It is a simple index of weight-for-height that is commonly used to classify underweight, overweight and obesity. It is defined as the weight in kilograms divided by the square of the height in meters. The international cut off points for child and adolescent according to age and sex as recommended by Cole et al. (2000, 2007) was computed to assess the classification of BMI.

Data on physical activity was collected using a structured schedule. Physical activity was divided into sedentary, moderate, and vigorous physical activity. Watching television or video or sitting in front of the computer daily were classified under sedentary behavior; household chores, moderately unorganized games and sports, walking to and from school were classified under moderate physical activity. And activities such as physical exercise (running, jogging, etc.), vigorously unorganized games and sports (football, basketball, etc.) were classified under vigorous physical activity, that made one sweat or breathes hard for at least 20 minutes as mentioned in Biddle et al., (2004). No physically challenged, qualified athlete or sportsperson were included in the sample under study.

For the statistical analysis, t-test and chi-square test were employed to test the level of significance at 0.05 levels. All the data were analyzed and managed using Windows Microsoft Excel and Statistical Package for Social Sciences (SPSS) 17.

Result

Table 11.1 and Figure 11.1 and 11.2 reveals that there was a statistically significant difference in mean height between boys and girls during the age groups of 12 years ($t=2.09, p<0.05$) and 14 to 18 years ($t=2.90$, 8.01, 7.81, 8.82, 7.85 at $p<0.05$ respectively). Girls were found to be taller for 8 to 13 years and the boys became significantly taller thereafter. As for weight, it was noticed among 12 ($t=2.08, p<0.05$), 13 ($t=3.60, p<0.05$), and 16 to 18 years ($t=2.64$, 3.76, 3.12 $p<0.05$ respectively). Girls were heavier during 10 to 14 years and the boys became taller thereafter. Succinctly, sex differences according to height and weight were significantly noticed after 12 years of age.

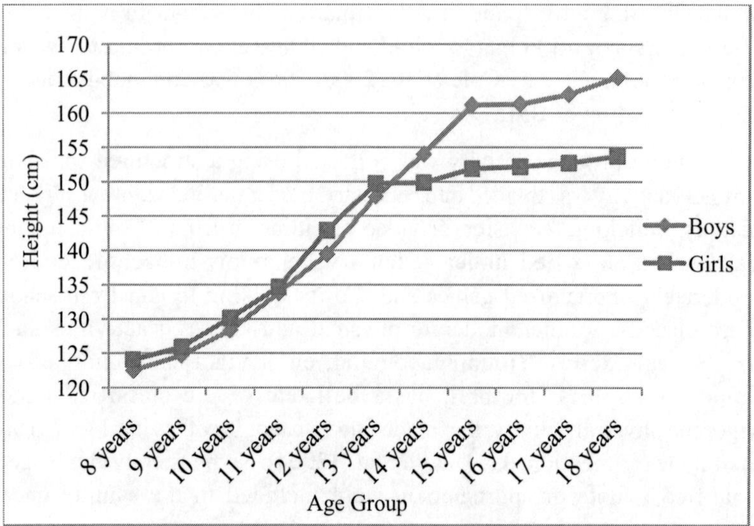

Fig. 11.1: Showing mean height of boys and girls according to age

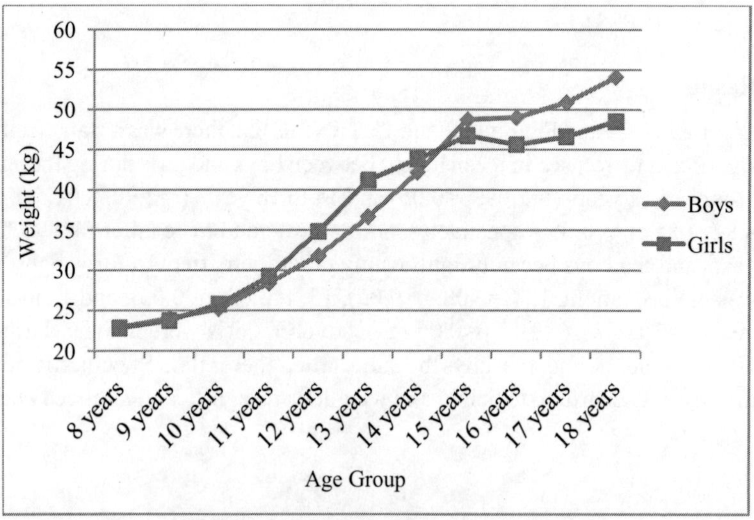

Fig. 11.2: showing mean weight of boys and girls according to age

Table 11.1: Age-wise between sex comparison (t-value) for mean height and weight

Age in years	*Height*	*Weight*
8	1.34	0.12
9	0.86	0.28
10	1.37	0.63
11	0.55	0.53
12	2.09*	2.08*
13	1.26	3.60*
14	2.90*	1.06
15	8.01*	1.39
16	7.81*	2.64*
17	8.82*	3.76*
18	7.85*	3.12*

* significant at 0.05 level.

As can be seen on Table 11.2, there was a significant difference in height among boys between the age groups 9 and 10 (t=2.68, p<0.05), 10 and 11 (t=3.58, p<0.05), 11 and12 (t=3.16, p<0.05), 12 and 13 (t=4.59, p<0.05), 13 and 14 (t=3.45, p<0.05), and 14 and 15 (t=4.75, p<0.05) years. Significant difference was also noticed in weight between the age groups 10 and 11 (t=2.63, p<0.05), 11 and 12 (t=2.26, p<0.05), 12 and 13 (t=3.74, p<0.05), 13 and 14 (t=3.94, p<0.05), 14 and 15 (t=3.80, p<0.05), and 17 and 18 (t=2.35, p<0.05) years. Girls also showed significant difference in height between the age groups 10 and 11 (t=3.34, p<0.05), 11 and 12 (t=5.81, p<0.05), and 12 and 13 (t=5.23, p<0.05) years. As for weight in girls, it was noticed between 9 and 10 (t=2.03, p<0.05), 10 and 11 (t=2.56, p<0.05), 11 and 12 (t=3.76, p<0.05), 12 and 13 (t=4.45, p<0.05), and 14 and 15 (t=2.39, p<0.05) years. Succinctly, age-wise comparison with regard to height and weight showed significant difference during 9 to 15 years in both boys and girls.

Figure 11.3 and 11.4 presents the distribution of BMI according to age in boys and girls respectively. Underweight was found to be prevalent in all the age groups and was higher among the younger age groups. On the other hand, overweight was prevalent at 9 to 12, 15, and 18 years of age in boys (Fig.11.3) and 9, 11, 13 to 16, and 18 years in girls (Fig.11.4). Overall, the prevalence of underweight was 31.44 percent in boys and 25.13 percent in a girl and overweight 1.73 percent in boys and 3.11 percent in girls. The

Table 11.2: Age-wise comparison (t-value) of height and weight in boys and girls

Age group in years	*Boys*		*Girls*	
	Height	*Weight*	*Height*	*Weight*
8-9	1.67	1.27	1.37	1.11
9-10	2.68*	1.22	1.37	2.03*
10-11	3.58*	2.63*	3.34*	2.56*
11-12	3.16*	2.26*	5.81*	3.76*
12-13	4.59*	3.74*	5.23*	4.45*
13-14	3.45*	3.94*	0.11	1.82
14-15	4.75*	3.80*	1.90	2.39*
15-16	0.11	0.17	0.22	0.88
16-17	1.34	1.66	0.44	0.77
17-18	1.94	2.35*	0.76	1.15

* Significant at 0.05 levels

distribution of BMI by age also showed a significant difference (χ^2=39.69, p<0.05) among girls (table not shown).

Descriptive statistics show that (Table 11.3), a higher percentage of the boys were vigorously active (53.22%), 26.49 percent sedentary, and 20.30 percent moderately active. Among girls, 46.63 percent were sedentary, 35.23 percent moderately active and 18.13 percent vigorously active. A statistical analysis between the children's physical activity level and BMI showed a significant association in the distribution of underweight

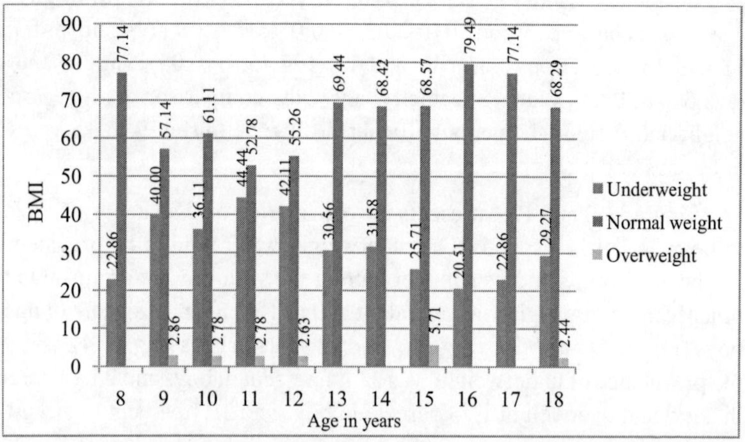

Fig.11. 3: Distribution of BMI by age in boys

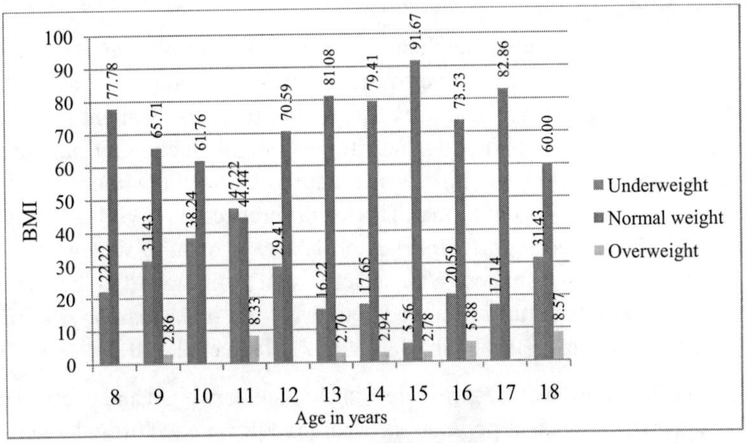

Fig. 11.4: Distribution of BMI by age in girls

(χ^2=19.22, p<0.05 in boys and 14.90, p<0.05 in girls) and normal weight (χ^2=14.06, p<0.05 in boys and 8.29, p<0.05 in girls) according to activity level. A notably higher percentage of underweight boys (40.93%) and girls (38.57%) were vigorously active, and a higher percentage of boys (76.83%) and girls (78.33%) with normal weight were moderately active and sedentary respectively. And although not statistically profound, a higher percentage of overweight boys (3.74%) and girls (5.00%) were found to be sedentary and lowest among those who were vigorously active (0.47% in boys).

Table 11.3: Distribution of BMI according to physical activity in boys and girls

Physical Activity	*N*	*BMI*		
		Underweight N(%)	*Normal N(%)*	*Overweight N(%)*
Boys				
Sedentary	107(26.49)	22 (20.56)	81 (75.70)	4 (3.74)
Moderate	82(20.30)	17 (20.73)	63 (76.83)	2 (2.44)
Vigorous	215(53.22)	88 (40.93)	126 (58.60)	1 (0.47)
χ2 value		19.22*, df=2	14.06*, df=2	4.80, df=2
Girls				
Sedentary	180(46.63)	30 (16.67)	141 (78.33)	9 (5.00)
Moderate	136(35.23)	40 (29.41)	93 (68.38)	3 (2.21)
Vigorous	70(18.13)	27 (38.57)	43 (61.43)	0 (0.00)
χ^2 value		14.90*, df=2	8.29*, df=2	4.75, df=2

*significant at 0.05 level

Discussion

The increase in mean height and weight with the advancement of age, which is in consonance with various other studies (Widiyani et al., 2011, Dkhar and Singh, 2012), may be attributed to the improvement in socio-economic conditions and better nutrition. A slight fluctuating pattern in body weight at 15 years in girls is also interpreted with the understanding that the data were cross-sectional. The sex differences vis-à-vis body height and weight existed significantly after 12 years of age, which was in agreement with other studies (Widiyani et al., 2011). In support of the present study, several other studies showed that girls were taller than boys during younger age groups (Bhasin and Jain, 2007; Mann et al., 2011).

The distribution of BMI revealed that both underweight and overweight coexisted in the present population, where the prevalence of underweight was significantly higher. The problem of undernutrition among children and adolescents is a serious public health problem all over the world, more specifically, in developing countries. The high percentage of underweight children in the present study has been attributed to the imbalance between nutrients needed for proper growth and regular energy expenditure through physical activity. Another reason that may be taken into account is that most of the children in the present study belonged to families from the low-income group (51.73% in boys and 59.07% in girls). A similar result was interpreted in several other works (Widiyani et al., 2011; Maken and Varte, 2012; Wolde et al., 2015). Population from lower socio-economic status not only had poor access to food but also poor environmental conditions such as poor housing and hygienic conditions, unsafe drinking water, heavy workloads, lack of preventive and control measures of locally endemic diseases and infections which were also attributed to high prevalence of underweight (de Onis et al., 2000; Khongsdier, 2002). The prevalence of underweight boys was higher in boys (31.44%) than in girls (25.13%), and overweight higher in girls (3.11%) than in boys (1.73%) in the present study. Earlier onset of puberty and early menarche were found to be associated with higher levels of fatness and higher risk for overweight and obesity among girls. In contrast, boys ordinarily declined in relative fatness during puberty because of the rapid growth of the fat-free mass (Malina et al., 2004c). Additionally, the prevalence of overweight in the present study is also attributed to the reduction in energy expenditure accompanied by an increase in energy intake. Factors such as lifestyle and food habits can increase the amount of fat in the body including body weight, which is a phenomenon that is increasingly becoming common in developing country (Bodzsar et al., 2005; Laxmaiah et al., 2007).

Progressing sedentary behavior and missing regular exercise in childhood and adolescence affects unfavorably not merely the existing health status and mental or physical capacity of performance, but increases the risks of chronic adult diseases as well (Due et al., 2001). Studies have also revealed that sedentary behavior such as time spent in watching television was directly related to increases in BMI (Wu, 2006; Laxmaiah et al., 2007). A high percentage of the children in the present population was also found to spend a growing fraction of their free time in sedentary behavior such as watching television or videos or sitting in front of the computer on a daily basis.

Vigorous physical activity differs between boys and girls across ages. Boys are commonly found to be more active than girls, and the physical activity declined with age, more so among girls. This decline in physical activity level could be attributed to various biological changes associated with puberty and social demands. It should also be taken into consideration that much of the physical activities took place after school, and the diminishing activity may be attributed to topography, among many others, where the present area is hilly in nature, the streets are narrow, and there are only a few available playgrounds or space which indirectly affects the activity level.

Conclusion

It is clearly evident from the present study that the growth of the Ao Naga children increased with age and there were a significant age and sex differences. A significantly close association between nutritional status and physical activity was also noticed, and their relationship was compounded by various factors. The cause-and-effect relationship noticed between BMI and sedentary behavior in the present study is of considerable concern obviously because television influences physical activities. Vigorous physical activity declined with age more so among girls. And although studies have shown that low physical activity levels most frequently occur in advanced developed countries, it is becoming increasingly common in developing countries as well. Keeping in mind that underweight has been one of the significant problems in developing countries, overweight and obesity are also increasing rapidly in recent times, creating a double burden of both under and overweight. In the present studied population, although the prevalence of underweight was significantly higher, overweight was also not uncommon. It is therefore important to take preventive measures early in life, as the health problems early in life may have long-term consequences that may persist into adulthood. Intervention programs should be implemented not only to prevent the increase in overweight and obesity but also to eradicate the age-old undernutrition and its associated morbidities, especially among the urban poor. As children spend a high percentage of their waking hours at school, schools can become the ideal platform and key environment to educate children about the importance of active living and healthy diet that would benefit their knowledge, attitudes, and behavior.

References

Bhasin, MK, and Jain, S 2007, 'Biology of the tribal groups of Rajasthan, India: 6. A comparative study of the nutritional status'. *Anthropologist*, Vol. 9, no.4, pp. 273-279.

Biddle, SJH, Gorely, T and Stensel, DJ 2004, 'Health-enhancing physical activity and sedentary behaviour in children and adolescents'. *Journal of Sports Sciences*, Vol. 22, no.8, pp. 679-701.

Bodzsar, EB, Zsakai, A, Jakab, K and Toth, KB 2005, 'Body fatness and sexual maturation status'. *Anthropological Notebooks* (Szlovenia), Vol. X, no.1, pp. 12-17.

Cole, TJ, Bellizzi, MC, Flegal, KM and Dietz, WH 2000, 'Establishing a standard definition for child overweight and obesity worldwide: International survey'. *British Medical Journal*, Vol. 320, pp. 1240-1243.

Cole, TJ, Flegal, KM, Nicholls, D, and Jackson, AA 2007, 'Body mass index cut offs to define thinness in children and adolescents: International survey'. *British Medical Journal*, Vol. 335, pp. 194-197.

De Onis, M, Frongillo, EA, and Blossner, M 2000, 'Is malnutrition declining? An analysis of changes in levels of child malnutrition since 1980'. *Bulletin of the World Health Organisation*, Vol. 78, pp. 1222-1233.

Dkhar, JW, and Singh, MS 2012, 'Prevalence of overweight and obesity among the Meitei adolescent of Manipur, India'. *Anthropologist*, Vol. 14, no.4, pp. 347-351.

Due, P, Hickman, M and Komkov, A 2001, Physical activity. Currie, C, Samdal, O, Boyce, W and Smith, R (eds.), *Health behaviour in school-aged children: a WHO cross-national study (HBSC). Research protocol for the 2001/2002 survey*. Child and Adolescent Health Research Unit (CAHRU), University: Edinburgh, pp. 59-70.

Khongsdier, R 2002, 'Body mass index and morbidity in adult males of the War Khasi in Northeast India'. *European Journal of Clinical Nutrition*, Vol. 56, pp. 484-489.

Laxmaiah, A, Nagalla, B, Vijayaraghavan, K and Nair, M 2007, 'Factors affecting the prevalence of overweight among 12 to 17-year-old urban adolescents in Hyderabad, India'. *Obesity*, Vol. 15, no.6, pp. 1384-1390.

Maken, T and Varte, LR 2012, 'Nutritional status of children as indicated by z-scores of the Hmars: A tribe of N.E. India'. *Journal of Anthropology*, Vol. 8, no.1, pp. 1973-1980.

Malina, RM and Little, BB 2008, 'Physical activity: The present in the context of the past'. *American Journal of Human Biology*, Vol. 20, pp. 373-391.

Malina, RM, Bouchard, C and Bar-Or, O 2004a, *Physical activity as a factor in growth, maturation, and performance. In: Growth, maturation, and physical activity*. Second edition. USA: Human Kinetics, pp. 479-508.

Malina, RM, Bouchard, C and Bar-Or, O 2004b, *Obesity in childhood and adolescence. In: Growth, maturation, and physical activity*. Second edition. USA: Human Kinetics, pp. 527-551.

Malina, RM, Bouchard, C and Bar-Or, O 2004c, *Other factors affecting growth, maturation, performance, and activity. In: Growth, maturation, and physical activity*. Second edition. USA: Human Kinetics, pp. 553-578.

Mann, PK, De, D, Bera, TK, Chatterjee, K and Gosh, D 2011, 'Anthropometric assessment of physical growth and nutritional status among school children of North Bengal'. *Anthropologist*, Vol. 13, no.4, pp. 299-305.

Martin, R and Saller, K 1957, *Lehrbuch der anthropologie*. Gustav Fischer Verlag, Stuttgart.

Snedecor, GW, and Cochran, WG 1967, *Statistical methods. Iowa*. The Iowa State University Press.

Widiyani, T, Suryobroto, B, Budiarti, S and Hartana, A 2011, 'The growth of body size and somatotype of Javanese children age 4 to 20 years'. *HAYATI Journal of Biosciences*, Vol. 18, no.4, pp. 182-192.

World Health Organisation 1997, *Global strategy for non-communicable disease prevention and control (draft)*. Geneva, Switzerland: WHO.

Wolde, M, Berhan, Y and Chala, A 2015, 'Determinants of underweight, stunting and wasting among schoolchildren'. *BMC Public Health*, Vol. 15, no.8.

Wu, YF 2006, 'Overweight and obesity in China'. *British Medical Journal*, Vol. 333, pp. 362-363.

Chapter - 12

Anthropometric Profile and Nutritional Status of an Elderly Tribal Population in Northeast India

— S. Khamu and B.T. Langstieh

Introduction

The rapid and fast technological progress made in medical science has controlled the fertility and mortality rates considerably and the average composition exhibits a relatively larger proportion of the elderly person. This demographic transition is a major force shaping the 21st century. The proportion of elderly persons in India has risen from 5.6 percent in 1961 to 7.4 percent in 2001 and is expected to be 12.4 percent in 2026 (MSPI, 2011). In 2004, the 60th round of the National Sample Survey Organization (NSSO) studied the condition and health care of the aged. The survey estimated that in the rural areas, every 1000 persons in the working age have to provide support, physically or otherwise, to 125 aged persons (old age dependency ratio). Among states, the proportion of elderly persons vary from around 4 percent in small states like Dadra and Nagar Haveli, Nagaland, Arunachal Pradesh, Meghalaya to more than 10.5 percent in Kerala (MSPI, 2011).

Anthropometry is the single most portable, universally applicable, inexpensive and non-invasive method available to assess the proportions, size, and composition of the human body. It reflects both health and nutritional status and may be used to predict performance, health, and survival (WHO, 1995). For these reasons, it is used for selecting individuals and populations for health and nutritional interventions, as well as for monitoring their health and nutrition (de Onis and Habicht, 1996). Anthropometric evaluation is an essential feature of geriatric nutritional evaluation and is highly reliable for determining malnutrition, being overweight, obesity, muscular mass loss, fat mass gain and adipose tissue redistribution (Perissinotto et al., 2002; Sanchez-García et al., 2007) when compared with more sophisticated methodologies, the use of which is restricted by complexity and cost in population studies.

In order to determine the possibility or risk of developing disease complications, indicators such as Waist Circumference (WC), Waist to Hip Ratio (WHR) and Body Mass Index (BMI) or Quetelet's Index are often used as indicators or measure of the health of a person, and also to identify individuals with the risk of developing serious health conditions, such as cardiovascular diseases, obesity and hypertension in type 2 diabetes. WC is considered a good estimate of the body fat, especially the internal fat deposits and the likelihood of developing weight-related diseases, while, WHR is a simple anthropometric index for assessing the amount and distribution of body fat (Al-Ansari and Mokdad, 2015).

Body Mass Index is often used for the purpose of nutritional surveillance. It is a simple but objective anthropometric indicator of the nutritional status of the adult population (Shetty and James, 1994). It has been recognized that there are variations in the relationship between BMI, body fatness and morbidity and mortality in different ethnic groups. There is strong evidence that metabolic disease occurs at lower BMI in Asian populations, which might be explained by a greater propensity to store abdominal fat in these populations (Caterson and Gill, 2002). Therefore, there has been an attempt to interpret the WHO BMI cut-offs using different criteria for Asian and Pacific populations based on risk factors and morbidities (WHO, 2000). However, WHO, 2004 maintains that these cut-offs are not significantly different and, therefore, the standard WHO BMI classifications should be used in Asian, African, American and Polynesian people.

Materials and Methods

The State of Nagaland was formally inaugurated on December 1st, 1963 as the 16th state of the Indian union and constitutes one of the seven States in the Northeastern part of India. The State approximately lies between 25°6' and 27°4' latitude, north of the equator and between the longitudinal lines 93°20'E and 95°15'E. The 2011 Census of India report recorded the total population as 1,978,502 and it was found to be the only State in the country to record a decrease in population with - 0.58 percent decadal growth since 2001. The State consists of eleven administrative districts, inhabited by 16 major tribes along with other sub-tribes. Chakhesang constitutes one of the major tribes. Each tribe is distinct in character from the other in terms of customs, languages, and dresses.

For the present study, the sample was drawn from the Chakhesang population inhabiting the rural areas, comprising both men and women who were 60 years or over. Since the study was conducted in rural areas, it is to be noted that the main occupation is farming. The data collection was done in 20 Chakhesang villages under Phek district, Nagaland, viz., Enhulumi, Chizami, Kami, Khezhakeno, K. Basa, K. Bawe, Lekromi, Lasumi, Leshemi, Losami, Mesulumi, Pfutseromi, Phusachodu, Phuyoba, Rihuba, Sakraba, Thenyizu, Thetsumi, Thipuzu, and Zapami. No statistical sampling technique was applied for the selection of the sites because the criterion was dependent on the presence or absence of elderly participants in any of the villages. Purposive sampling was taken into consideration to include those elders who were 60 years and above, who were mentally receptive and willing to participate in the study. Most aged individuals did not have an official record of their birth date; as such, the age was estimated by matching recall of particular historical events. The arrival of the Japanese army to the area during World War II (April 1944) was most often used as the point of reference.

A schedule was used for gathering individual data on socio-demographic parameters such as the name of the subject, date, and place at which data was recorded, sex, age, education, occupation, type of income, marital status, and living arrangement.

The anthropometric measurements were taken and recorded in accordance with the techniques described by Weiner and Lourie (1981). Martin's anthropometer, sliding caliper, weighing machine, flexible non-elastic measuring tape, and Harpenden's skin fold caliper were used. Measurements were taken on the left side because it is less likely to be affected by various factors, e.g., occupational deformity, change due to extra work, etc. (Das and Deka, 2013). The length, breadth and circumferential measurements were recorded to the nearest 1.0 mm while skin fold measurement was recorded to the nearest 0.1 mm.

The weight of the subject was taken by means of a standard weighing machine. The subject was instructed not to wear any footwear, headgear, etc., and wear light garments.

Height measurements were taken for stature in Frankfurt Horizontal plane and knee-heel length/knee height. The subjects were made to stand erect without any kind of footwear, with the feet running parallel to one another, heels touching one another, arms hanging maximum on the sides, palms touching the thighs and the head resting on the eye-ear or the Frankfurt Horizontal plane without any kind of footwear. The horizontal bar is lowered to touch the crown of the head in the mid-sagittal plane with sufficient pressure to compress the hair. Knee height is the measurement predicated only on adults 60 years of age and older. To obtain the measurement, the subject sits with legs dangling. The fixed socket with the crossbar is placed under the heel of the left leg just below the lateral malleolus of the fibula. With the knee and ankle at a 90° angle, the socket with the adjustable crossbar is placed on the anterior surface of the left thigh, above the condyles of the femur while resting the subject's foot on the palm of the investigator.

The breadth measurements included knee breadth and elbow breadth. Knee breadth was measured with the subject seated and the knee flexed at 90°. With a blunt-end spreading caliper, measurement was taken from between the most bilateral and most medial bony protrusions of the knee below the mid-level of the patella, taking care not to measure the femur. For measuring the elbow breadth, the left arm was flexed so that the elbow forms a 90° angle with the fingers pointing up and the posterior part of the wrist toward the investigator. With the sliding caliper held at a 45° angle to the plane of the long axis of the upper arm, the greatest breadth across the epicondyles of the elbow was measured.

The body circumferences that were measured included elbow amplitude, mid-upper arm circumference (MUAC), calf circumference, hip circumference (HC) and waist circumference. For measuring elbow amplitude, the measuring tape was placed around the elbow, with the tape resting on the skin surface while taking care not to compress the skin. The MUAC and calf circumference were measured around the mid-upper arm at the marked point perpendicular to the long axis of the upper arm and around the largest circumference of the calf while the subject was seated with the knee and ankle bent at a 90° angle. The HC was measured with the tape placed at the maximum extension of the hip making sure that that the tape is in a horizontal plane. The WC was taken around the trunk and at the plane between the umbilical scar and the inferior rib border. The WC was used to identify individuals with possible health risks especially increased risk of metabolic complications, based on cut-offs for mene" 90 cm and women ≥ 80 cm specifically meant for South Asian populations (IDF, 2006).

Skinfold measurement was taken around the triceps, biceps, calf, thigh, subscapular, and suprailiac. For measuring the triceps and biceps, the midpoint on the posterior surface of the left upper arm and the skin fold at the midpoint on the anterior aspect of the upper arm (10 mm above the level used to mark the triceps site) respectively are picked up and measured while holding the skin fold. The calf skinfold was measured on the inside of the right leg at the level of maximal calf girth while the thigh skinfold was measured in the midline of the anterior aspect of the right thigh. For the subscapular skinfold measurement, the scapula was palpated to locate the inferior angle and then picking up the skin fold just under the shoulder blade following the natural fold of the skin. The suprailiac skinfold was measured by determining the mid-axillary line and palpated for the iliac crest (top of the hip bone) with the skin fold sloping downward at a 45° angle extending toward the pubic symphysis.

The BMI was categorized according to WHO, 2004 standards and it was estimated by dividing weight (kg) by height^2 (m^2). Individuals were considered malnourished if their BMI was less than 18.5, normal from 18.5 to 24.9 and overweight if it is ≥ 25.

WHR was estimated by dividing waist circumference by hip circumference. The threshold WHR was ≥ 0.85 for women and ≥ 0.90 for men (ICMR, 2005), above which a superior distribution of adipose tissue was considered.

Data were managed using the Statistical Package for the Social Sciences (SPSS) windows software, version 16. Further, anthropometric data were presented as mean and SD. Chi-square test was performed to look at the levels of association with regard to socio-demographic characteristics between genders. ANOVA was performed to identify age differences or similarities of anthropometric measurements in men and women and within the entire sample. The comparison between means across gender was performed by Student's two-tailed t-test for independent data.

Results

Table 12.1 presents the socio-demographic characteristics of the Chakhesang elders across gender. The study was carried out among 912 elders, with 425 men (46.6%) and 487 women (53.4%). The age range of the participants ranged from 60-101 years with the mean age, M= 72.6 years. The nuclear family type was the most common type of living arrangement, while the majority of the elders were reportedly still in wedlock. Majority of the elders were illiterate or had a low educational level, and cultivators. As with monetary income from Government sources, 32.7 percent were recipients of the National Old Age Pension Scheme. The rest were solely dependent upon farming, husbandry, basketry, weaving, etc. for generating income. There were significant differences (p<0.05) between men and women in the living arrangements, marital status, educational and occupational levels, and levels of activity. In comparison to men, women were significantly more likely to live alone, widowed, less educated and to have low occupational status. On the other hand, men were more likely to be married than women. The most common level of activity is the primary level which includes everyday agricultural chores. Both men and women are equally involved in agricultural activities as there is no concept of 'conventional housewife' in this society. More men are involved in the secondary level of activity such as axing, lifting heavy loads, clearing forests, etc. A comparatively low percentage of the elders are sedentary,

Table 12.1: Socio-demographic characteristics of the Chakhesang elders

Characteristics		*Men*		*Women*		*Total*	
		N	*%*	*N*	*%*	*N*	*%*
Sample Size (N)		425	46.6	487	53.4	912	100
Age (Mean) in years		73.6		71.8		72.6	
Living Arrangements*							
Nuclear		164	38.6	166	34.1	330	36.2
Joint		124	29.2	134	27.5	258	28.3
Living with a spouse		89	20.9	69	14.2	158	17.3
Living alone		48	11.3	118	24.2	166	18.2
Marital Status*							
Married		333	78.4	252	51.7	585	64.1
Widowed		70	16.5	212	43.5	282	30.9
Unmarried		21	4.9	18	3.7	39	4.3
Divorced		1	0.2	5	1.0	6	0.7
Educational Levels*							
Illiterate		259	60.9	453	93.0	712	78.1
Primary		75	17.6	18	3.7	93	10.2
Middle		50	11.8	11	2.3	61	6.7
High School		33	7.8	5	1.0	38	4.2
Higher Secondary		5	1.2	-	-	5	0.5
Graduate		3	0.7	-	-	3	0.3
Occupational Levels*							
Farmers		336	79.1	477	97.9	813	89.1
In Govt. Service		14	3.3	3	0.6	17	1.9
Retired Govt. Servants		61	14.4	6	1.2	67	7.3
Religious		4	0.9	1	0.2	5	0.5
Village Officials		10	2.4	-	-	10	1.1
Levels of Activity*							
Primary		252	59.3	342	70.2	594	65.1
Secondary		153	36.0	122	25.1	275	30.2
Sedentary		20	4.7	23	4.7	43	4.7

*p<0.05

mainly due to infirmities or sicknesses. However, those elders were also actively involved in farming in the past.

With an aim to render an anthropometric profile of an elderly population in Northeast India, an attempt was made to develop an anthropometric data set on the elderly Chakhesang population. At present, very less anthropological investigations have been done among elderly persons in this part of the country and there are no published data on the anthropometric measurements of elderly Chakhesang people.

Anthropometric values according to age and gender are presented in Table 12.2. Anthropometric measurements showed that the age-group statistics differed in terms of MUAC, calf circumference, stature, weight, triceps skinfold, biceps skinfold, suprailiac skin fold, and sub-scapular skin fold ($p<0.05$). Age-group statistics among the men differed in terms of MUAC, calf circumference, stature, and elbow amplitude ($p<0.05$). While, among women, age-group statistics differed in terms of MUAC, calf circumference, WC, stature, weight, triceps skinfold, biceps skinfold, calf skinfold, thigh skin fold, and subscapular skinfold ($p>0.05$). While, no age-group statistics were found to differ in terms of HC, knee-heel length, knee breadth, and elbow breadth in the present sample ($p>0.05$).

A comparison between gender shows that men tend to have greater MUAC, calf circumference, WC, stature, weight, knee-heel length, knee breadth, elbow breadth, elbow amplitude, thigh skin fold and subscapular skinfold ($p<0.05$), while values for skin fold measurements such as triceps, biceps, calf, and suprailiac skinfold were greater among the women ($p<0.05$). Surprisingly, men and women did not differ in terms of HC in the present study ($p<0.05$).

For each age group, men had greater MUAC, calf circumference, stature, weight, knee-heel length, knee-breadth, elbow breadth, and elbow amplitude than women. No significant difference between genders was observed for HC in any of the age-groups.

BMI was also used to determine underweight and overweight (Table 12.3). None of the subjects in the present study belonged to the 'obese' category as defined by WHO, 2004. It was found that 21.3 percent of the elders were underweight, while a low 6.2 percent of the elders were found to be overweight. Majority of the elders (72.5 %) were found to belong to the normal category. Being underweight was observed among 14.0 percent of the 60-64 years age group, while among the 90+ age group, the percentage

Table 12.2: Anthropometric values according to age and gender

	Men		*Women*		*Total*			*Men*		*Women*		*Total*	
	Mean	*SD*	*Mean*	*SD*	*Mean*	*SD*		*Mean*	*SD*	*Mean*	*SD*	*Mean*	*SD*
Mid Upper Arm Circumference abc							***Calf Circum ference abc***						
60-64d	24.7	2.1	23.4	1.9	23.9	2.1	**60-64d**	34.0	2.7	31.6	2.3	32.5	2.7
65-69d	24.2	2.1	23.2	2.2	23.6	2.2	**65-69d**	33.1	2.7	30.9	2.4	31.9	2.7
70-74d	24.0	2.0	22.5	2.1	23.2	2.2	**70-74d**	33.3	3.2	30.3	2.7	31.7	3.3
75-79d	23.8	2.0	22.8	1.9	23.3	2.0	**75-79d**	33.2	2.7	30.2	2.9	31.7	3.1
80-84d	23.6	2.0	22.2	1.8	22.9	2.0	**80-84d**	32.4	2.9	29.7	2.4	31.0	3.0
85-89d	23.9	2.1	22.2	2.0	23.0	2.2	**85-89d**	33.4	2.9	29.9	3.0	31.6	3.4
90+	22.4	1.9	21.7	1.9	22.2	1.9	**90+d**	31.1	2.4	28.8	2.8	30.2	2.7
Total	24.0	2.1	22.8	2.0	23.4	2.1	**Total**	33.1	2.9	30.5	2.6	31.7	3.0
Waist Circumference							**Hip Circumference**						
60-64d	77.6	9.7	74.0	8.1	75.4	8.9	**60-64**	87.4	5.0	86.1	4.8	86.6	4.9
65-69d	76.9	8.8	74.1	7.7	75.3	8.3	**65-69**	86.3	5.5	85.8	5.0	86.0	5.2
70-74d	76.8	9.4	72.7	8.0	74.7	8.9	**70-74**	86.0	5.4	84.6	4.7	85.2	5.1
75-79	76.8	9.1	74.8	8.7	75.8	8.9	**75-79**	85.6	5.5	85.9	5.4	85.8	5.4
80-84	76.2	6.6	76.4	8.7	76.3	7.7	**80-84**	85.9	5.0	85.5	4.3	85.7	4.7
85-89	78.5	9.9	76.3	8.7	77.3	9.3	**85-89**	86.5	6.1	85.6	4.6	86.0	5.4
90+	76.0	6.4	78.8	10.4	77.1	8.1	**90+**	85.6	3.8	86.1	4.0	85.8	3.8
Total	77.0	8.8	74.5	8.3	75.6	8.6	**Total**	86.2	5.3	85.6	4.8	85.9	5.0
Stature abc							**Weight bc**						
60-64d	162.4	5.7	151.7	5.4	156.0	7.6	**60-64d**	55.5	8.3	47.7	6.3	50.8	8.1
65-69d	160.9	6.0	149.6	5.4	154.7	8.0	**65-69d**	54.0	8.6	46.2	6.4	49.7	8.4
70-74d	159.9	6.8	149.2	5.2	154.2	8.0	**70-74d**	53.8	8.6	44.3	6.5	48.8	8.9
75-79d	161.1	6.0	147.9	5.7	154.4	8.8	**75-79d**	53.1	8.3	45.0	6.8	49.0	8.6
80-84d	157.9	5.8	148.0	6.1	153.0	7.7	**80-84d**	51.8	7.4	44.7	5.1	48.2	7.3
85-89d	159.6	5.9	147.4	6.4	153.3	8.7	**85-89d**	53.2	7.8	44.5	6.0	48.8	8.2
90+d	157.4	8.5	145.2	4.4	152.8	9.3	**90+d**	49.1	7.8	40.8	5.6	46.0	8.0
Total	160.3	6.4	149.3	5.7	154.4	8.2	**Total**	53.5	8.3	45.5	6.4	49.2	8.3

(Contd...)

Knee-Heel Length							Knee Breadth						
60-64d	49.0	2.8	45.7	2.5	47.1	3.1	**60-64d**	9.3	0.6	8.5	0.6	8.8	0.7
65-69d	49.1	2.3	45.6	2.5	47.2	3.0	**65-69d**	9.3	0.6	8.5	0.4	8.9	0.6
70-74d	48.8	2.9	46.0	2.5	47.3	3.0	**70-74d**	9.4	0.5	8.4	0.6	8.9	0.7
75-79d	49.5	2.4	45.7	2.4	47.5	3.0	**75-79d**	9.4	0.4	8.5	0.4	8.9	0.6
80-84d	49.0	2.8	46.4	2.2	47.7	2.8	**80-84d**	9.3	0.4	8.4	0.4	8.8	0.6
85-89d	49.8	2.4	46.4	3.0	48.0	3.2	**85-89d**	9.3	0.9	8.5	0.5	8.9	0.8
90+d	49.5	2.8	46.8	2.6	48.5	3.0	**90+d**	9.2	0.4	8.3	0.4	8.9	0.6
Total	49.1	2.6	45.9	2.5	47.4	3.0	**Total**	9.3	0.5	8.5	0.5	8.9	0.7
Elbow Breadth							**Elbow amplitude A**						
60-64d	6.7	0.4	6.0	0.4	6.3	0.5	**60-64d**	24.3	1.9	22.1	1.1	23.0	1.8
65-69d	6.8	0.5	5.9	0.3	6.3	0.6	**65-69d**	23.9	1.6	22.1	1.9	22.9	2.0
70-74d	6.7	0.4	6.0	0.5	6.3	0.6	**70-74d**	23.7	1.4	21.7	1.3	22.7	1.7
75-79d	6.7	0.4	6.0	0.6	6.4	0.6	**75-79d**	23.9	1.3	22.0	1.3	22.9	1.6
80-84d	6.6	0.4	6.0	0.5	6.3	0.5	**80-84d**	23.5	1.4	21.7	1.8	22.6	1.8
85-89d	6.6	0.3	6.0	0.4	6.3	0.5	**85-89d**	23.6	1.7	22.4	3.1	23.0	2.6
90+d	6.8	0.3	6.2	0.5	6.6	0.5	**90+d**	23.0	1.2	21.2	1.1	22.3	1.5
Total	6.7	0.4	6.0	0.4	6.3	0.5	**Total**	23.8	1.6	22.0	1.7	22.8	1.9
Triceps Skinfold bc							**Biceps Skin-fold bc**						
60-64d	5.1	2.4	7.9	3.1	6.8	3.2	**60-64d**	2.7	1.1	3.3	1.4	3.0	1.3
65-69d	4.6	1.7	7.8	3.3	6.4	3.2	**65-69d**	2.5	0.9	2.9	1.2	2.8	1.1
70-74d	4.4	1.7	6.3	2.7	5.4	2.5	**70-74**	2.5	1.0	2.8	1.3	2.7	1.1
75-79d	4.3	1.8	6.6	2.6	5.5	2.5	**75-79**	2.5	0.8	2.9	1.5	2.7	1.2
80-84d	4.6	1.7	6.1	2.3	5.3	2.2	**80-84**	2.6	1.0	2.7	0.9	2.7	1.0
85-89	4.9	2.3	5.6	2.6	5.3	2.5	**85-89**	2.7	1.1	2.5	1.0	2.6	1.0
90+	4.6	1.3	5.5	2.6	4.9	1.9	**90+**	2.4	0.6	2.1	1.1	2.3	0.8
Total	4.6	1.9	6.9	3.0	5.8	2.8	**Total**	2.6	0.9	2.9	1.3	2.7	1.1

(Contd...)

Calf skinfold b							Thigh skinfold b						
60-64d	4.6	1.9	5.6	2.5	5.2	2.3	**60-64d**	5.0	1.9	4.1	1.2	4.5	1.6
65-69d	4.5	1.6	5.3	2.4	4.9	2.1	**65-69**	4.6	1.3	4.4	1.4	4.5	1.3
70-74	4.7	1.9	4.6	2.0	4.6	1.9	**70-74d**	4.9	1.4	3.9	1.3	4.3	1.5
75-79	4.6	2.6	4.6	2.2	4.6	2.4	**75-79d**	4.6	1.4	3.8	1.2	4.2	1.3
80-84	4.8	1.8	4.8	2.3	4.8	2.1	**80-84d**	4.8	1.3	3.8	1.0	4.3	1.2
85-89d	5.4	2.2	4.2	2.4	4.8	2.4	**85-89d**	4.9	1.1	3.7	1.1	4.3	1.2
90+	4.5	1.5	6.4	3.8	5.2	2.7	**90+**	4.3	0.8	4.3	2.1	4.3	1.4
Total	4.7	2.0	5.0	2.4	4.9	2.2	**Total**	4.8	1.4	4.0	1.3	4.4	1.4
Suprailiac skinfold							**Subscapular skinfold bc**						
60-64d	7.8	4.7	9.3	5.2	8.7	5.1	**60-64**	8.0	3.2	7.1	2.4	7.5	2.7
65-69d	6.9	3.4	9.2	5.0	8.2	4.5	**65-69d**	7.5	3.0	6.5	2.1	7.0	2.6
70-74	7.1	3.4	7.9	4.0	7.5	3.7	**70-74d**	7.2	2.3	6.1	2.1	6.6	2.3
75-79d	7.6	4.5	9.5	5.0	8.6	4.8	**75-79**	6.9	2.1	6.5	1.9	6.7	2.0
80-84	7.2	4.2	7.5	3.4	7.4	3.8	**80-84**	6.7	1.9	6.4	2.2	6.5	2.1
85-89	7.4	4.4	8.6	5.2	8.0	4.8	**85-89d**	6.9	2.6	5.9	1.6	6.4	2.2
90+d	6.1	1.7	9.2	3.1	7.3	2.7	**90+**	6.9	1.8	6.3	1.5	6.7	1.7
Total	7.2	4.0	8.7	4.7	8.0	4.4	**Total**	7.2	2.6	6.5	2.1	6.8	2.4
BMI							**WHR bc**						
60-64	21.0	2.9	20.7	2.4	20.9	2.6	**60-64d**	0.9	0.1	0.9	0.1	0.9	0.1
65-69	20.8	2.8	20.6	2.5	20.7	2.6	**65-69d**	0.9	0.1	0.9	0.1	0.9	0.1
70-74d	21.0	2.6	19.9	2.6	20.4	2.6	**70-74**	0.9	0.1	0.9	0.1	0.9	0.1
75-79	20.5	2.9	20.5	2.5	20.5	2.7	**75-79**	0.9	0.1	0.9	0.1	0.9	0.1
80-84	20.7	2.3	20.4	2.5	20.6	2.4	**80-84**	0.9	0.1	0.9	0.1	0.9	0.1
85-89	20.9	2.6	20.5	2.3	20.7	2.4	**85-89**	0.9	0.1	0.9	0.1	0.9	0.1
90+	19.8	2.1	19.3	2.2	19.6	2.1	**90+**	0.9	0.1	0.9	0.1	0.9	0.1
Total	20.8	2.7	20.4	2.5	20.6	2.6	**Total**	0.9	0.1	0.9	0.1	0.9	0.1

(a) Significant differences between age range in women ($p < 0.05$), (b) Significant differences between age range in men ($p < 0.05$), (c) Significant differences between age range ($p < 0.05$) (d) Significant differences between women and men ($p < 0.05$).

Table 12.3: Body Mass Index (BMI) according to age and gender

BMI		*60-64*	*65-69*	*70-74*	*75-79*	*80-84*	*85-89*	*90+*	*Total*
Men									
<18.5	N	9	15	19	15	8	6	6	78
	%	13.0	16.9	21.8	24.6	12.5	16.2	33.3	18.4
18.5 – 24.9	N	54	66	60	42	53	29	12	316
	%	78.3	74.2	69.0	68.9	82.8	78.4	66.7	74.4
≥ 25-29	N	6	8	8	4	3	2	-	31
	%	8.7	9.0	9.2	6.6	4.7	5.4	-	7.3
Women									
<18.5	N	15	23	37	13	15	8	5	116
	%	14.6	21.1	38.1	20.3	23.4	20.5	45.5	23.8
18.5 – 24.9	N	82	79	56	48	45	29	6	345
	%	79.6	72.5	57.7	75.0	70.3	74.4	54.5	70.8
≥ 25-29	N	6	7	4	3	4	2	-	26
	%	5.8	6.4	4.1	4.7	6.2	5.1	-	5.3
Total									
<18.5	N	24	38	56	28	23	14	11	194
	%	14.0	19.2	30.4	22.4	18.0	18.4	37.9	21.3
18.5 – 24.9	N	136	145	116	90	98	58	18	661
	%	79.1	73.2	63.0	72.0	76.6	76.3	62.1	72.5
≥ 25-29	N	12	15	12	7	7	4	-	57
	%	7.0	7.6	6.5	5.6	5.5	5.3	-	6.2

was higher at 37.9 percent. Whereas, being overweight was not observed among the 90+-year-olds.

In the present study, the indicators such as WC and WHR were used as indicators of the health of a person, and also to identify individuals with the risk of developing serious health conditions. With regard to WC, it was found that 9.6 percent men and 24.8 percent of elderly women belonged to the 'at risk' category (WC ≥ 90 and WC ≥ 80 cm) as defined by IDF (2006). For every age group, men were found to be relatively at a lesser percentage of risk. Furthermore, with WHR cut-offs, 41.6 percent of men (WHR ≥ 0.9) and 57.1 percent women (WHR ≥ 0.85) belonged to the 'at risk' category. The percentages implied towards a higher percentage of women with central adipose tissue distribution and likely risks than men and the trend was observed in all the age groups.

Table 12.4: Waist Circumference and Waist to Hip Ratio as predictors of disease risk factors according to age and gender

		60-64	*65-69*	*70-74*	*75-79*	*80-84*	*85-89*	*90+*	*Total*
WC									
Men									
≥ 90 cm	N	8	10	10	6	1	5	1	41
	%	11.6	11.2	11.5	9.8	1.6	13.5	5.6	9.6
<90cm	N	61	79	77	55	63	32	17	384
	%	88.4	88.8	88.5	90.2	98.4	86.5	94.4	90.4
Women									
≥ 80 cm	N	23	23	17	17	20	15	6	121
	%	22.3	21.1	17.5	26.6	31.2	38.5	54.5	24.8
<80cm	N	80	86	80	47	44	24	5	366
	%	77.7	78.9	82.5	73.4	68.8	61.5	45.5	75.2
WHR									
Men									
≥0.9	N	25	37	35	27	27	18	8	177
	%	36.2	41.6	40.2	44.3	42.2	48.6	44.4	41.6
<0.9	N	44	52	52	34	37	19	10	248
	%	63.8	58.4	59.8	55.7	57.8	51.4	55.6	58.4
Women									
≥ 0.85	N	50	64	49	36	44	26	9	278
	%	48.5	58.7	50.5	56.2	68.8	66.7	81.8	57.1
<0.85	N	53	45	48	28	20	13	2	209
	%	51.5	41.3	49.5	43.8	31.2	33.3	18.2	42.9

Discussion

In the present cross-sectional study, we attempted to present an anthropometric profile of an elderly population and tried to compare our findings with those provided by similar studies. However, the lack of anthropometric cross-sectional surveys in the Northeastern part of India has limited the comparison of our gender and age-wise results with those other studies.

Gender differences were observed in stature and body weight. Men were significantly (p≥0.05) taller (160.3 versus 149.3) and heavier (53.5 versus 45.5). The results are consistent with those carried out by Kuczmarski et al., 2000; Lorini et al., 2014; Perissinotto et al., 2002; Sanchez-Garcia et al., 2007; Santos et al., 2004. The decrease in height and weight was observed as age increased; however, the decreases were not consistent with increasing age groups. Other studies also found that height and weight decreased with ageing (Corish et al., 2003; Odenigbo et al., 2010; Rosnah et al., 2009).

Table 12.5: Comparison of Stature and Weight with other Studies

Reference	*Nationality*	*Gender*	*Age Limits*		*N*	*Stature*	*SD*	*Weight*	*SD*
			Lower	*Upper*					
Al-Ansari and Mokdad., 2015	Bahrain	Men	60	92	34	165.0	9.2	73.56	14.69
		Women	60	92	50	152.6	5.4	80.03	18.0
Barbosa, et al., 2005	Brazil	Men	60	+	770	164.4	0.06	68.20	12.41
		Women	60	+	1124	150.0	0.05	62.57	0.06
Faruque et al., 2006	Bangladesh	Men	60	106	718	159.7	6.1	47.7	8.6
		Women	60	106	478	147.0	5.7	41.1	9.4
Gavriilidou et al., 2015	Sweden	Men	60	99	1454	175.4	7.6	84.4	14.1
		Women	60	99	1688	161.4	6.7	70.8	20.2
Hu et al., 2007	China	Men	65.2	85.1	50	165.5	54.3	68	10.6
		Women	65.0	80.7	55	152.6	69.3	60	9.7
Kothiyal and Tettey, 2000	Australia	Men	65	92	33	165.8	7.9	72	11
		Women	65	92	138	152.1	-	61	-
Perissinotto et al., 2002	Italy	Men	65	84	5462	171.7	-	72.6	-
		Women	65	84	5462	152.2	-	63.8	-
Rosnah et al., 2009	Malaysia	Men	60	+	129	162.3	7.5	66.6	11.3
		Women	60	+	101	149.0	5.8	60.0	13.8
Sánchez-García et al., 2007	Mexico	Men	60	+	1098	163.2	8.5	70.3	12.3
		Women	60	+	870	152.6	7.5	62.7	11.6
Santos et al., 2004	Chile	Men	60	99	411	164.6	7.1	73.2	13.0
		Women	60	99	807	149.8	6.3	63.6	13.4
Present Study	**India**	**Men**	**60**	**101**	**450**	**160.3**	**6.4**	**53.5**	**8.3**
		Women	**60**	**101**	**462**	**149.3**	**5.7**	**45.5**	**6.4**

Table 12.5 shows the comparison of stature and weight with that of studies from other nations. It is evident that older individuals from the developed nations (Australia, Italy, Bahrain, and Sweden) are taller than those from developing nations (Chile, Brazil, Bangladesh, China, Mexico, and Malaysia).

BMI values were significantly different by gender. While most studies show women to have higher mean BMI (as women tend to have more fat deposits) than that of men from similar ages (Odenigbo et al., 2010; Perissinotto et al., 2002; Santos et al., 2004), the present study found that men had a higher mean BMI than the women (Table 12.4). BMI represents the easier and most frequently used index to identify subjects at risk for under or over nutrition. While BMI appears to have excellent validity as a measure of absolute fat mass adjusted for height in young and middle-aged adults, it has its limitations in detecting the physiological age-related transformations of lean mass to fat mass in older adults (WHO, 2008; Zamboni et al., 2005). BMI does not distinguish between weight due to fat mass and muscle mass; and this can lead to misclassifying individuals (e.g. athletes and well-trained bodybuilders) and populations (e.g., Australian Aborigines) who differ in body build and body proportions (extremes of age and height; Caterson and Gill, 2002), and is an inaccurate method to estimate body fatness in elderly people aged 75+ years because of body composition changes such as a reduction in skeletal muscle mass and an increase in visceral fat deposits in the abdomen that occur with ageing (Espinel and King, 2009). Many authors agree in considering this index a poor indicator of risk in the elderly (Allison et al., 1997; Sánchez-García et al., 2007; Seidell and Visscher, 2000), because it does not reflect regional distribution of fat or any change in fat distribution in the elderly (Perissinotto et al., 2002). The same adult BMI value corresponds to more fatty body composition in the elderly. Thus, the BMI has to be differently interpreted for elderly subjects.

The most obvious changes associated with ageing concern body composition with declination in muscle mass which is gradually replaced by fat mass (Babiarczyk and Turbiarz, 2012). Furthermore, fat location changes over time, with fat mass tending to increase around the abdomen as we age, which can often lead to serious metabolic consequences (Ahmed and Haboubi, 2010; Goulart and Rexrode, 2007; Villareal et al., 2005). Hence, alternative direct and indirect methods to measure body fat distribution have emerged and are now widely used including waist circumference and waist-to-hip ratio (Espinel and King, 2009). For the elderly people, waist circumference and waist-to-hip ratio correlate better with blood pressure and lipid profile than BMI and have been suggested as better indicators for overall body fatness (Goya et al., 2004). Although many techniques are available to assess the visceral fat deposits, the simplest and the most clinically useful measurement is the waist circumference (Snijder et al., 2006). In the present study, BMI was found to significantly and positively correlate with both WC ($r=0.76, p \geq 0.05$) and WHR ($r=0.47, p \geq 0.05$), however, the correlation was strongly correlated with WC compared to WHR. Espinel and King (2009) have also opined that WC relates closely to BMI and is considered an appropriate indicator for use in monitoring weight status among adults at the population level. Similar studies which show waist circumference as a better predictor of adiposity than BMI in older adults have also been reported by NSW Health, (2003); Visscher et al., (2001); Zamboni et al., (2005).

There are no existing evidence-based practice guidelines to assist clinicians in classifying BMI for elderly populations. However, there is strong emerging evidence that WHO cut-offs may not be appropriate in increasing age (Queensland Government, 2014). In the absence of an ideal BMI cut-off for the elderly population, there have been suggestions by some authors that BMI thresholds for malnutrition should be modified. Sergi et al., 2005 suggested the threshold at < 20.0, while the Queensland Government has suggested adjusting BMI classifications for underweight people aged ≥ 65 years the threshold at < 23kg/m^2. If in case that value is applied to the present population under study, the categorisation for underweight or the malnourished category according to Sergi et al., 2005 and Queensland Government will be 43.9 percent and 84.2 percent respectively as against 21.3 percent according to WHO (2004) classification, while, categorising underweight which are likely overrated and an exaggerated percentages.

While obesity is a major problem which is plaguing most western countries, developing countries continue to confront with underweight and undernutrition. The present study shows a lower mean BMI than most reported studies (Table 12.6) with a higher percentage of elders belonging to the underweight category. This is likely due to the regular and heavy physical activities among the rural populace understudy, who are wholly engaged in farming. However, on the positive side, physical activity engagement is considered an important component of lifestyle definition. There are suggestions that engaging in physical activity might alleviate degenerative body compositional changes (Dunsky et al., 2014). In a study of 1,663 Israeli older adults, Dunsky et al. (2014) found that the more active a person was, the lower the values of the parameters and therefore the negative association between weight and BMI and physical activity is probably a result of reduced fat, which is obviously more desirable than reduced muscle mass. Numerous other studies (Brunet et al., 2007; Dey et al., 2001; Dunsky et al., 2014; Kyle et al., 2006; Leskinen et al., 2009; Tsai et al., 2007; Waller et al., 2008) have also confirmed negative associations between level of physical activity and weight, and WC and BMI values in both genders. Willett (1997) has suggested that diminished lean body mass (contributing to weight loss) leads to poor health, whereas decreased fat mass (also leading to weight loss) would, in contrast, be associated with improved health. On the other side of the scale, undernutrition is a well-known predictor of mortality (Visscher et al., 2000), and modification of cut-off values to identify patients at nutritional risk has been suggested (Corish et al., 2003). Diet and physical activity modulate ageing associated anthropometric changes (Li et al., 2011; Reddy and Papa, 2010; Marais et al., 2007; Price et al., 2007). In a study of older adults in Jinja, Cheserek (2012) found that the older adults had higher lean body mass and less fat, thus more protein reserves, but clearly deficient fat stores, which was attributed to limitations in caloric adequacy in contrast to protein intake, or to intensive physical activity. This suggests that, in developing countries, older adults may be equally vulnerable to protein deficiency (ibid).

Table 12.6: Comparison of BMI with other Studies

BMI	*Nationality*	*Age Limit*		*Men*		*Women*		*Underweight*		*Overweight*	
				Mean	*SD*	*Mean*	*SD*	*Men %*	*Women %*	*Men %*	*Women %*
Ansari and Mokdad, 2015	Bahrain	60	92	26.9	4.2	34.2	6.7	-	-	-	-
Chen et al., 2012	Malaysia	60	97	22.3	4.0	23.5	5.5	15.4	20.7	24.2	35.6
Cheserek et al., 2012	East Africa	60	+	20.2	3.3	21.6	4.7	29.5	24.2	8.4	12.5
Gavriilidou et al., 2015	Sweden	60	99	25.8	3.7	25.3	4.6	-	-	-	-
Sánchez-García et al., 2007	Mexico	60	+	26.4	4.4	26.8	4.4	1.2	1.6	59.9	65.4
Present Study	India	60	101	20.8	2.7	20.4	2.5	18.4	23.8	7.3	5.3

After heavy physical exertions, it is often recommended that a high-quality protein source (like meat, fish or poultry, dairy, eggs, and legumes like beans and lentils) combined with a carbohydrate meal be consumed to help maintain the body's protein balance, more particularly for older adults. However, the consumption of protein-rich diet was lacking among the studied population. Hence, in order to cope with the process of ageing and also to compensate for the heavy physical activities, it is important for the older people to eat high-quality protein foods as proteins are sources of energy and are vital for the maintenance of body tissue, development, and repair. Timely intervention can stop weight loss in the elderly at risk of malnutrition.

However, it is hoped that the findings of the present study would have shed some light on the anthropometric details and also the high prevalence of underweight elderly. In the present study, older individuals seem to have low-fat mass. BMI has been believed to overestimate the prevalence of overweight individuals (Sánchez-García et al., 2007), however, considering the current population under study, BMI can also lead to an overestimation of underweight individuals, who may otherwise appear to be apparently healthy individuals. In this case, BMI may not be very viable and efficient in estimating underweight individuals. Hence, in addition to the fact that BMI may not be a very reliable index for estimation of nutritional status among the elderly, estimation of BMI among the very physically active and older individuals (who lack in fat mass but not lean mass) may face a double disadvantage. The limitation of the present study was that it did not try to evaluate the lean body weight/fat-free mass from fat mass. Further in-depth research on the accurate estimation for underweight and overweight among older individuals needs to be considered, taking into consideration other factors such as diet and body composition changes, social and health conditions, as well as by employing more sophisticated techniques including an estimate of body compartments (fat-free mass and fat mass) by instrumental methods such as bioelectrical impedance analysis and dual X-ray absorptiometry, as suggested by Enzi et al., 1997, to arrive at a more accurate evaluation of nutritional status. Moreover, reference values for the evaluation of the nutritional status in older adults (like BMI) have been based upon extrapolations on studies using young adults or based on statistical definitions of threshold values, rather than on the elder's morbidity, mortality and quality of life (Sánchez-García et al., 2007). Therefore, it is necessary to consider all of these factors to determine desirable threshold values for anthropometric measures in the elderly population.

References

Ahmed, T and Haboubi, N 2010, 'Assessment and management of nutrition in older people and its importance to health'. *Clinical Interventions in Ageing,* Vol. 5, pp. 207-216.

Al-Ansari, MS and Mokdad, M 2015, 'Elderly anthropometrics for ergonomic purposes'. *Journal of Educational Psychological Science,* Vol. 16, pp. 605-632.

Allison, DB, Gallagher, D, Heo, M, Pi-Sunyer, FX and Heymsfield, SB 1997, 'Body mass index and all-cause mortality among people age 70 and over: The longitudinal study of ageing'. *International Journal of Obesity,* Vol. 21, pp. 424-431.

Babiarczyk, B, and Turbiarz, A 2012, 'Body mass index in elderly people - Do the reference, ranges matter?' *Progress in Health. Science,* Vol. 2, pp. 58-67.

Brunet, M, Chaput, J, and Tremblay, A 2007, 'The association between low physical fitness and high body mass index or waist circumference is increasing with age in children: the quebec en forme project'. *International Journal of Obesity,* Vol. 31, pp. 637-643.

Caterson, ID and Gill, TP 2002, 'Obesity: Epidemiology and possible prevention'. *Best Practice and Research Clinical Endocrinology Metabolism,* Vol. 16, pp. 595-610.

Census of India 2011, *Census of India. Registrar General and Census Commissioner of India, Delhi*. http://www.census2011.co.in/census/state/nagaland.html.

Cheserek, MJ, Tuitoek, PJ, Waudo, JN, Msuya, JM and Kikafunda, JK 2012, 'Anthropometric characteristics and nutritional status of older adults in the Lake Victoria Basin of East Africa: Region, sex, and age differences'. *South African Journal of Clinical Nutrition,* Vol. 25, pp. 67-72.

Corish, CA, Flood, P, Mulligan, S and Kennedy, NP 2003, 'Apparent low frequency of undernutrition in Dublin hospital in-patients: Should we review the anthropometric thresholds for clinical practice?' *British Journal of Nutrition*, Vol. 84, pp. 325-335.

Das, BM, and Deka, R 2013, *Physical anthropology practical*. Kitab Mahal, Allahabad.

De Onis, M and Habicht, J 1996, 'Anthropometric reference data for international use: Recommendations from a world health organization expert committee'. *American Journal of Clinical Nutrition*, Vol. 64, pp. 650-658.

Dey, DK, Rothenberg, E, Sundh, V, Bosaeus, I and Steen, B 2001, 'Height and body weight in elderly adults: A 21-year population study on secular trends and related factors in 70-year olds'. *Journal of Gerontology, A. Biological Science and Medical Science*, Vol. 56, pp. M780-M784.

Dunsky, A, Zach, S, Zeev, A, Goldbourt, U, Shimony, T, Goldsmith, R and Yael, Netz 2014, 'Level of physical activity and anthropometric characteristics in old age-Results from a national health survey'. *European Review of Ageing and Physical Activity*, Vol. 11, pp. 149-157.

Enzi, G, Sergi, G, Bussolotto, M, Ceccon, A, Giantin, V and Beninca, P 1997, Methods for quantitation of body composition, with particular reference to lean body mass. Tessari, P, Soeters, PB, Pittoni, G and Tiengo, A (Eds.) *Amino Acid and Protein Metabolism in Health and Disease*. Smith-Gordon, Great Britain, pp. 5-11.

Espinel, PT, and King, L 2009, *A framework for monitoring overweight and obesity in NSW.* NSW Department of Health and Physical Activity. Nutrition Obesity Research Group, Sydney.

Goulart, AC, and Rexrode, KM 2007, 'Health consequences of obesity in the elderly: A review'. *Current Cardiovascular Risk Reports,* Vol. 1, pp. 340-347.

Goya, WS, Gerald, SA, Whincup, PH and Walker, M 2004, 'Overweight and obesity and the burden of disease and disability in elderly men'. *International Journal of Obesity and Related Metabolic Disorders,* Vol. 28, pp. 1374-1382.

Indian Council of Medical Research 2005, *Guidelines for management of type 2 diabetes.* ICMR, New Delhi.

International Diabetes Federation 2006, *The IDF consensus worldwide definition of the metabolic syndrome.* IDF, Brussels.

Kuczmarski, MF, Kuczmarski, RJ and Najjar, M 2000, 'Descriptive anthropometric reference data for older Americans'. *Journal of American Dietetic Association,* Vol. 100, pp. 59-66.

Kyle, UG, Melzer, K, Kayser, B, Picard-Kossovsky, M, Gremion, G and Pichard, C 2006, 'Eight-year longitudinal changes in body composition in healthy Swiss adults'. *Journal of the American College of Nutrition,* Vol. 25, pp. 493-501.

Leskinen, T, Sipila, S, Alen, M, Cheng, S, Pietilainen, KH, Usenius, JP, Suominen, H, Kovanen, V, Kainulainen, H, Kaprio, J and Kujala, UM 2009, 'Leisure-time physical activity and high-risk fat: A longitudinal population-based twin study'. *International Journal of Obesity,* Vol. 33, pp. 1211-1218.

Li, CL, Chen, SY, Lan, C, Pan, WH, Chou, HC, Bai, YB, Tzeng, MS, Lee, MS and Lai, JS 2011, 'The effects of physical activity, body mass index (BMI) and waist circumference (WC) on glucose intolerance in older people: A nationwide study from Taiwan'. *Archives of Gerontology and Geriatrics,* Vol. 52, pp. 54-59.

Lorini, C, Collini, F, Castagnoli, M, Di Bari, M, Cavallini, MC, Zaffarana, N, Pepe, P, Lucenteforte, E, Vannacci, A and Bonaccorsi, G 2014, 'Using alternative or direct anthropometric measurements to assess risk for malnutrition in nursing homes'. *Nutrition,* Vol. 30, pp. 1171-1176.

Marais, ML, Marais, D and Labadarios, D 2007, 'Assessment of nutritional status of older people in homes for the aged in the Somerset West Area'. *South African Journal of Clinical Nutrition,* Vol. 20, pp. 102-108.

Ministry of Statistics and Programme Implementation 2011, a *Situation analysis of the elderly in India*. The government of India, New Delhi.

National Sample Survey Organisation 2004, *Morbidity, health care and the condition of the aged*, 60th round, Ministry of Statistics and Programme Implementation. The government of India, New Delhi.

NSW Department of Health 2003, *Report on the weight status on NSW: 2003*, Centre for Public Health Nutrition, Sydney.

Odenigbo, UM, Odenigbo, CU and Oguejio for, OC 2010, 'Nutritional status of elderly in Asaba, Delta State, Nigeria'. *Pakistan Journal of Nutrition,* Vol. 9, pp. 416-418.

Perissinotto, E, Pisent, C, Sergi, G, Grigoletto, F and Enzi, G 2002, 'Anthropometric measurements in the elderly: Age and gender differences'. *British Journal of Nutrition*, Vol. 87, pp. 177-186.

Price, GM, Uauy, R, Breeze, E, Bulpitt, CJ and Fletcher, AE 2006, 'Weight, shape, and mortality risk in older persons: elevated waist-hip ratio, not high body mass index, is associated with a greater risk of death'. *American Journal of Clinical Nutrition*, Vol. 84, pp. 449-460.

Queensland Government 2014, *Using body mass index. A consensus document from Dietitian/ Nutritionists from the nutrition education materials* Online, "NEMO", team. http://www.health.qld.gov.au/masters/copyright.asp.

Reddy, KK, and Papa, R 2010, 'Nutritional status impair functional ability among the elderly'. *Journal of Anthropology,,* Vol. 3, pp. 192-199.

Rosnah, MY, Mohd Rizal, H, and Sharifah Norazizan, SAR 2009, 'Anthropometry dimensions of older Malaysians: Comparison of age, gender, and ethnicity'. *Asian Social Science,* Vol. 5, pp. 133-140.

Sánchez-García, S, García-Pena, C, Duque-Lopez, MX, Juarez-Cedillo, T, Cortes-Nunez, AR and Reyes-Beaman, S 2007, 'Anthropometric measures and nutritional status in a healthy elderly population'. *BMC Public Health,* Vol. 7, pp. 1-9.

Santos, JL, Albala, C, Lera, L, Garci, C, Arroyo, P, Pe'rez-Bravo, F, Angel, B and Pela'ez, M 2004, Anthropometric measurements in the elderly population of Santiago, Chile. *Nutrition,* Vol. 20, pp. 452-457.

Seidell, JC and Visscher, TLS 2000, 'Body weight and weight change and their health implications for elderly'. *Eurasian Journal of Clinical Nutrition,,* Vol. 54, pp. S33-S39.

Sergi, G, Perissinotto, E, Pisent, C, Buja, A, Maggi, S, Coin, A, Grigoletto, F, Enzi, G and ILSA Working Group 2005, 'An adequate threshold for body mass index to detect underweight condition in elderly persons: The Italian longitudinal study on ageing'. *Journal of Gerontology, A. Biological Science and Medical Science*, Vol. 60, pp. 866-871.

Shetty, PS, and James, WPT 1994, *Body mass index: A measure of chronic energy deficiency in adults.* United Nations, Rome.

Snijder, MB, van Dam, RM, Visser, M and Seidell, JC 2006, 'What aspects of body fat are particularly hazardous and how do we measure them'? *International Journal of Epidemiology,* Vol. 35, pp. 83-92.

Tsai, A, Liou, J, Chang, M and Chuang, Y 2007, 'Influence of diet and physical activity on ageing associated body fatness and anthropometric changes in older Taiwanese'. *Nutrition Research,* Vol. 27, pp. 245-251.

Villareal, DT, Apovian, CM, Kushner, RF and Klein, S 2005, 'Obesity in older adults: Technical review and position statement of the American society for nutrition and NAASO, the obesity society'. *American Journal of Clinical Nutrition,* Vol. 82, pp. 923-934.

Visscher, TL, Seidell, JC, Molarius, A, van der Kuip, D, Hofman, A and Witteman, JC 2001, 'A comparison of body mass index, waist-hip ratio and waist circumference as predictors of all-cause mortality among the elderly: The rotterdam study'. *International Journal of Obesity and Related Metabolic Disorders,* Vol. 25, pp. 1730-1735.

Visscher, TLS, Seidell, JC, Menotti, A, Blackburn, H, Nissinen, A, Feskens, EJM and Kromhout, D 2000, 'Underweight and overweight in relation to mortality among men aged 40–59 and 50–69 Years'. *American Journal of Epidemiology,* Vol. 151, pp. 660-666.

Waller, A, Kaprio, J and Kujala, UM 2008, 'Associations between long-term physical activity, waist circumference, and weight gain: A 30- year longitudinal twin study'. *International Journal of Obesity,* Vol. 32, pp. 353-361.

Weiner, JS, and Lourie, JA 1981, *Practical human biology*. Academic Press, London.

Willett, WC 1997, 'Weight loss in the elderly: Cause or effect of poor health'? *American Journal of Clinical Nutrition,* Vol. 66, pp. 737-738.

World Health Organization 1995, *Physical status: The use and interpretation of anthropometry*. WHO, Geneva.

World Health Organization 2000, *The Asia Pacific perspective: Redefining obesity and its treatment*. International Diabetes Institute. WHO, Australia.

World Health Organization 2004, 'Appropriate body-mass index for Asian populations and its implications for policy and intervention strategies: WHO expert consultation'. *Lancet,* Vol. 363, pp. 157-163.

World Health Organization 2008, *Waist circumference and waist-hip ratio: Report of a WHO expert consultation*. WHO, Geneva.

Zamboni, M, Mazzali, G, Zoico, E, Harris, TB, Meigs, JB, Di Francesco, V, Fantin, F, Bissoli, L and Bosello, O 2005, 'Health consequences of obesity in the elderly: A review of four unresolved questions'. *International Journal of Obesity,* (Lond). Vol. 29, pp. 1011-1029.

Chapter - 13

Association of Red Meat Consumption with Diabetes and Hyperglycaemia among the Urban Bhutia of Sikkim, India

— ***Sovanjan Sarkar***

Introduction

The World Health Organization (WHO) reported that about 80.0 percent of the deaths due to diabetes occur in middle-income countries (WHO, 2014). Cho et al. (2018) reported that in 2017 there were 451 million (age 18/years and above) people with diabetes worldwide which were expected to increase to approximately 700 million by 2045. It was also reported that there were an estimated 374 million people with impaired glucose tolerance (IGT) and approximately 5 million deaths worldwide were attributable to diabetes among adults. Type 2 diabetes mellitus or non-insulin dependent diabetes mellitus (NIDDM) is now noticed in almost every population and epidemiological evidence suggests that the prevalence will continue to increase globally if active preventive measures and control programmes were not taken inappropriate time (IDF, 2006). This report also estimated that approximately 80.00 percent of people with diabetes die of cardiovascular disease.

It has been found that every fifth diabetic in the world is an Indian. As per the International Diabetes Federation, the largest number of diabetic patients (50.8 million) resides in India for which, India continues to be the "diabetes capital" of the world, and by 2025, the expected number of diabetics in India is anticipated to be 68 million (IDF, 2006). Diabetes was previously prevalent only among adults. However, due to lifestyle change and adverse eating habits, hyperglycaemia is now found even among adolescents and children. As reported by the clinicians, diabetes is such a disease which can be managed but can't be cured. As a result of this increasing trend of diabetes in India, eventually, the prevalence of syndrome-X of metabolic syndrome is also increasing rapidly.

NIDDM is caused by the interaction of certain genetic and lifestyle factors (Kaprio et al., 1992). However, genes that predispose an individual to diabetes require the presence of environmental and behavioural factors, particularly those related to lifestyle. People with NIDDM have a multifold higher risk of coronary heart disease than the rest of the population. Moreover, the risk of cerebrovascular and peripheral vascular disease is also significantly higher among the NIDDM patient (Tuomilehto et al., 2004). Lifestyle factors like changes in diet, less physical activity, with consequent increases in the prevalence of overweight and obesity, have reported being associated with hyperglycaemia and NIDDM (Zimmet, 2001). Scientists from Harvard School of Public Health (HSPH) have reported that higher red meat consumption has a strong association with NIDDM (Harvard magazine, 2018). Several studies have reported the positive association of red meat consumption (both processed and unprocessed) with an increased risk of diabetes. After considering 12 cohort studies, Aune et al. (2009) suggested a positive association of red meat consumption with increased risk of diabetes. Another study reported an increased risk of coronary heart disease, stroke, and diabetes with higher consumption (Micha et al., 2010). Kurotani et al. (2013) also showed an increased risk of diabetes with an excess quantity of red meat intake among Japanese adults. Increased risk of diabetes, associated with higher red meat consumption, has been shown among the adults of different countries and multiethnic populations (Martin et al., 2012; Steinbrecher et al., 2010).

In India, the fast food culture and sedentary lifestyle are the two main drivers responsible for the diabetes epidemic (Mohan et al., 2007). Several studies have been done among different population groups of India pertaining to diabetes and hyperglycaemia and their lifestyle related correlates. However, to the best of author's knowledge, an association of red meat consumption with NIDDM was not been examined in any of the population groups in India and among the Northeastern tribal and non-tribal groups, exposed variously to modernizing forces. Hence, the objective of the present study is to assess the relationship between red meat consumption with diabetes and hyperglycaemia.

Materials and Methods

Study area

Gangtok, the capital of the sub-Himalayan State of Sikkim and its neighbourhood have been chosen as the study area.

Study populations

The Bhutias, a tribal population, migrated to Sikkim and other parts of India sometime after the 17th century (Risley, 1928). The Bhutias live in remote rural areas as well as in the capital town of Gangtok and its neighbourhood. They speak a language known as '*Sikkimese*'. However, they also speak in Nepali. Many Bhutias, especially the young educated segment can speak Hindi and English. Bhutias are traditionally rice eaters. They consume rice with animal fat, fried vegetables or meat (usually pork, and beef, and occasionally mutton or chicken). Consumption of beef and pork was very much predominant among them as they perceive that these would keep them warm in extremely cold climate. By religion they are Buddhist. *Chhyang* (traditional millet beer) is their favourite alcoholic drink and tea with salt is their preferred beverage. The other aspects of the sociodemographic profile of the Bhutias have been reported earlier (Sarkar and Mukhopadhyay, 2008).

Study participants

The present study has selected 340 Bhutias of both sexes aged 18 years and above from 100 systematically chosen houses from Gangtok and its neighbourhood. All the family members who qualify the age range of the present study have been taken into consideration.

Data types

A pretested questionnaire was used to collect data on socio-demographic aspects, such as educational and occupational status, monthly expenditure, age, and so on, of the participants. Data on perceived notion on daily activity pattern was collected and energy expenditure was derived from the 24-hour recall method. The perceived notion of physical activity was categorized under three heads, namely heavy, moderate and light. Food frequency questionnaire was used to collect data on a number of servings of different community specific food items on daily/weekly/monthly/yearly basis. A number of servings of various food items per month were used in the inferential statistical analyses. Information on daily/weekly alcohol consumption and tobacco use (smoking and chewing) was collected with the help of a questionnaire. Anthropometric measurements i.e., height (cm.), weight (kg), circumferences (cm.) and skinfold thicknesses were taken following Lohman et al., (1988). The body mass index (BMI) was computed as weight/stature^2 (kg/m^2). Waist-hip-ratio was also derived as waist circumference/hip circumference. Skinfold thickness was measured in four sites of the body, viz. biceps, triceps, supra-iliac and sub-scapular, and the sum of these four measurements was used in the inferential statistical analyses. Perceived stress was calculated following Levenstein et al. (1993). Bioelectrical impedance, i.e., percent body fat (%) (PBF) and fat mass (kg) (FM) were estimated using Omron body fat monitor HBF-300, manufactured by Omron Corporation, Japan.

Blood Glucose Estimation

A 12-hours fasting blood sample was collected from each study participant and the glucose level was estimated with the help of a glucose monitor (Accu-Check, Alpha) from whole blood. The use of a glucose monitor in estimating the blood glucose level has been reported elsewhere (ADA, 2010; Sacks et al., 2011). A value of $\geq$ 110 mg/dL was considered as hyperglycaemia (Alberti and Zimmet, 1998).

Ethical Issues

Informed consent was collected from each of the study participants prior to collecting any data.

All statistical analyses were done with the help of SPSS 16.0 for Windows software. Sex effect on fasting blood glucose value was not found through t-test and hence all the basic and inferential analyses were performed on total population. Descriptive statistics (mean, standard deviation and standard error) of fasting blood glucose were calculated, while the prevalence of hyperglycaemia and diabetes were estimated at the initial stage of analyses. Adjusted linear and logistic regression analyses were performed to find out the significant predictors of FBG and hyperglycaemia, respectively. Finally, receiver operating characteristic (ROC) curve and area under the curve were generated to identify the most significant predictor of hyperglycaemia. The significant level was fixed at ($p<0.05$).

Results

Table 13.1 shows the socio-economic profile of the study populations, by sex. Irrespective of sex, most of the study participants were found to be ever married. Majority of the study participants have attained their literacy level up to higher secondary. While government and non-government services were predominant among the males, the majority of the females were homemakers. Monthly expenditure of nearly 75.00 percent of the households is between below Rs. 15,000.

Table 13.1: Socio-economic profile of the study population, by sex

Variable		*Percentage*	
		Male (n=180)	*Female (n=160)*
Marital status			
	Ever married	88.0	11.0
	Unmarried	12.0	89.0
Literacy status			
	Illiterate	10.0	35.0
	Primary	09.0	11.0
	Up to Higher Secondary	62.0	48.0
	Others[#]	19.0	06.0
	Occupation		
	Service	80.0	21.0
	Homemaker	----	67.0
	Others [$]	20.0	12.0
Monthly household expenditure			
Upto Rs. 5000		01.0	
Rs. 5001 – Rs. 10,000		27.0	
Rs. 10,001 – Rs. 15,000		46.0	
Above Rs. 15,000		26.0	

[#] Graduate, postgraduate and professional; [$] Retired, unemployed, student, business.

Descriptive statistics (mean and SD) of different lifestyle-related variables have been shown in Table 13.2. A number of servings of different food items per month, energy expenditure, and perceived stress index do not show any significant sex difference in the study population.

Table 13.2: Descriptive statistics of lifestyle related variables, by sex

Variable	*Male (n=180)*	*Female (n=160)*	*Sig. (t-test)*
	Mean ± SD	*Mean ± SD*	
Cereals	67.28 ± 13.65	67.85 ± 13.73	p>0.05
Pulses	28.42 ± 22.69	25.65 ± 23.40	p>0.05
Tubers	39.83 ± 24.36	45.75 ± 20.59	p>0.05
Vegetables	95.22 ± 32.38	92.35 ± 23.76	p>0.05
Red meats	37.95 ± 22.48	34.26 ± 21.90	p>0.05
Lean meat	3.46 ± 4.93	3.37 ± 4.28	p>0.05
Eggs	10.23 ± 12.87	10.53 ± 13.80	p>0.05
Fish	2.59 ± 3.06	2.42 ± 2.87	p>0.05
Milk products	75.55 ± 30.32	80.05 ± 30.52	p>0.05
Fruits	16.80 ± 18.67	18.85 ± 22.22	p>0.05
Energy expenditure (KCal)	2416.87 ± 554.25	2142.18 ± 419.69	p>0.05
Perceived stress index	0.48 ± 0.06	0.48 ± 0.07	p>0.05

Table 13.3 shows the prevalence of substance use by means of alcohol and tobacco use. A significantly higher number of males was found to be a user of tobacco in the form of smoke and/or smokeless than their female counterparts. In the same line, a significantly higher number of males were consumers of alcoholic beverages in the form of foreign liquor and traditional millet beer.

Table 13.3: Prevalence of substance use by means of alcohol and tobacco use, by sex

Variable	*Male (n=180)*	*Female (n=160)*	*Sig. (test of equality of proportion)*
	%	%	
Alcohol consumer	25.0	10.0	P<0.01
Tobacco smoker	24.0	06.0	P<0.01
Tobacco chewer	37.0	13.0	P<0.001

Table 13.4 shows the mean, standard error of the mean and standard deviation (SD) of fasting blood glucose value of the study parti-cipants, pooled for sex. The mean is found to be below the cut-off value of hyperglycaemia and diabetes. A blood glucose value of more than 110 mg/dL was found among 31.50 percent of the total population (not shown in table).

Table 13.4: Descriptive statistics of fasting blood glucose, pooled for sex

Variable	*Mean*	*SEM*	*SD*
Fasting blood glucose (mg/dL)	102.25	1.34	24.73

Table 13.5 depicts the result of adjusted linear regression analysis using age-adjusted fasting blood glucose as dependent and other socio-economic, lifestyle and obesity-related variables as independent, pooled for sex. The only red meat was found to predict fasting blood glucose significantly (β=0.174; R^2=0.107).

Table 13.6 shows the result of adjusted multivariate logistic regression analysis using hyperglycaemia as dependent and other socio-economic, lifestyle and obesity-related variables as independent, pooled for sex. Risk of diabetes was significantly affected by percent body fat, WHR, and red meat consumption.

Table 13.5: Result of adjusted linear regression analysis using age fasting blood glucose as dependent and other socioeconomic, lifestyle and obesity-related variables as independent, pooled for sex

Variable	*Standardized Coefficients (β)*	*Sig.*	*R^2*
Education	.076	.267	0.107
Occupation	-.024	.749	
Energy Expenditure	-.009	.884	
Percent Fat	.160	.119	
Alcohol consumer	.023	.698	
Waist circumference	.221	.591	
Abdomen circumference	.125	.326	
Body Mass Index	.146	.483	
Waist hip ratio	.094	.436	
Sum of skinfolds	.193	.074	
Perceived stress index	.020	.828	
Red meat consumer	.174	.006	
Milk product consumer	.091	.269	

Table 13.6: Result of adjusted multivariate logistic regression analysis using diabetes status as dependent and other socioeconomic, lifestyle and obesity-related variables as independent, pooled for sex

Variable	*Wald statistics*	*Sig.*	*Exp(B)*	*95.0% C.I. for EXP(B)*	
				Lower	*Upper*
Education	.027	.868	1.053	.570	1.947
Occupation	.122	.727	1.182	.462	3.021
Energy expenditure	.562	.454	1.000	1.000	1.000
Percent fat	5.165	.023	1.080	1.011	1.155
Alcohol consumer	.038	.845	.946	.544	1.646
Waist circumference	.061	.805	.783	.113	5.449
Abdomen circumference	.024	.877	1.048	.579	1.894
Body Mass Index	.350	.554	1.357	.494	3.727
Waist Hip Ratio	3.713	.044	1.765	.990	3.145
Sum of skin folds	.164	.686	.902	.549	1.483
Perceived stress index	.912	.340	.256	.016	4.185
Red meat consumer	5.997	.014	.985	.973	.997
Milk product consumer	2.835	.092	1.009	.999	1.019

Figure 13.1 illustrates the ROC curve analysis using diabetes status as dependent and WHR, percent fat and red meat as independent variables. These independent variables were found to be significant in predicting diabetes status of the study participants as per the result of adjusted multivariate logistic regression analysis.

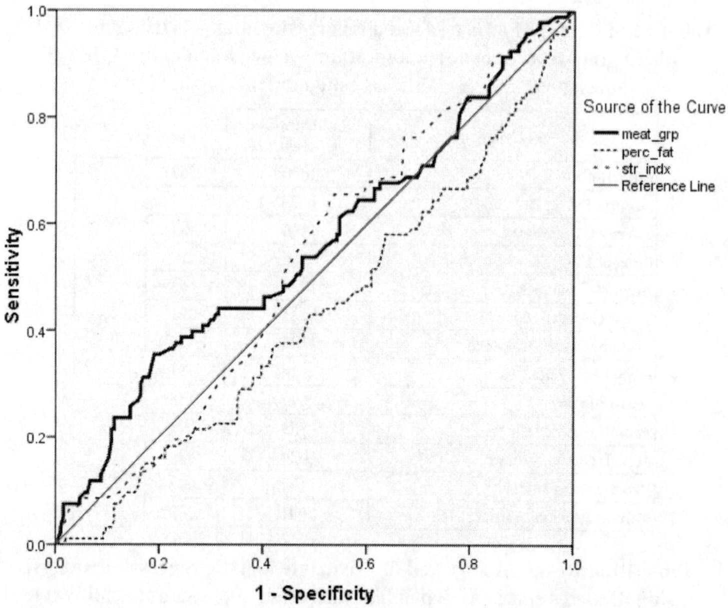

Figure 13.1: Receiver operating characteristic curve showing the area under the curve for red meat consumption, WHR and percent fat

The area under the curve of the ROC curve analysis has been shown in Table 13.7. It has been noticed that the area under the curve for red meat consumption was much higher (0.558) than other two body fat related variables (Percent fat 0.437 and WHR 0.489).

Table 13.7: Area under the curve for ROC curve analysis using red meat consumption, waist-hip ratio and percent body fat as predictors of diabetes status of the study participants

Area Under the Curve					
Test Result Variable(s)	*Area*	*Std. Error*[a]	*Asymptotic Sig.*[b]	*Asymptotic 95% Confidence Interval*	
				Lower Bound	*Upper Bound*
Red meat	.558	.037	.100	.486	.631
Percent fat	.437	.035	.075	.367	.506
WHR	.489	.036	.762	.418	.560

Discussion

The present small-scale cross-sectional study evidently demonstrated that higher red meat consumption increases the risk of high blood glucose and hyperglycaemia among the Bhutias of Gangtok. As per the declaration of WHO, in 2030, 80.00 percent of the total Indian population would be diabetic. Therefore, the time has arrived to implement certain awareness and treatment programmes to fight against the future death-defying scenario. And for implementation of such programmes, it is very much crucial to identify the lifestyle-related significant risk factor(s) of high blood glucose, hyperglycaemia, and diabetes, and also the prevalence of diabetes in different populations. Additionally, special attention should be given to those individuals, specifically adolescents and children, who are near the borderline of hyperglycaemia. Hence, this study has been done at an appropriate point in time to provide valuable information on the epidemiology of diabetes.

Bhutias are residing in the hilly terrain of Gangtok and its neighbourhood and are found to be under both lifestyle and epidemiological transition. The habitual physical activity level of the people was lower in comparison with their calorie intake. This may be due to a couple of reasons. Firstly, their socio-economic status which is comparatively higher than other population group residing in Gangtok. Being a Scheduled Tribe (ST), as declared by the Government of India, Bhutias are predominantly engaged in both government and non-government services and hence earning a handsome amount of salary per month. This helps them in availing hired vehicle to cover such distances which on the other way may be covered by foot. This, in turn, helps them accumulating extra fat in their body. Secondly, the easy availability of fat and fiber-rich junk food in the locality. Bhutias are fond of red meat in the form of beef and pork. They are regularly consuming those fast food made up of red meat and also found to purchase a high quantity of meat for their home for cooking. This double effect of consuming fast food and homemade food items with red meat put them in adverse situation pertaining to many chronic illnesses including blood glucose and hyperglycaemia. Consuming red meat in the home was their traditional habit. However, the changing food habit along with other behavioural factors may affect their health status in different ways. Perceived psychosocial stress may also play a responsible role as shown in an earlier study (Sarkar and Mukhopadhyay, 2008). It has also been observed that females are doing most of the household works. Playing cards and gambling along with consumption of liquor with snacks made up of red meat are the usual activity of the male Bhutias during leisure while watching a movie is the most favourite free time activity of the females. The present study didn't find any individual practicing physical exercise for health, neither anyone who is aware of the adverse effect of red meat consumption.

The mean value of fasting blood glucose and the prevalence of hyperglycaemia was found to be more than 100.0 mg/dL and more than 30.00 percent, respectively, which is immensely alarming for the study population. No sex difference has been observed in the mean value of FBG. Moreover, no sex difference is evident in any of the lifestyle-related variables like a number of servings of different food items, energy expenditure and perceived psychosocial stress.

The inferential statistical analyses were done on total population, pooled for sex, because the result of t-test did not show any significant effect sex on age-adjusted FBG. Among the food items, energy expenditure, body fat related measures, perceived stress and other socio-economic status variables, red meat found to be the only lifestyle related variable predicting age-adjusted fasting blood glucose significantly in linear regression analysis. Again, adjusted logistic regression analysis predicts WHR, red meat consumption and percent body fat as three significant predictors of hyperglycaemia. Except for red meat, this finding corroborated with another study (Dowse et al., 1991). To find out the best predictor out of these three significant correlates, the ROC curve was generated which finally depicts red meat consumption as the best predictor of hyperglycaemia in the study population. Effect of higher red meat consumption with an elevated level of blood glucose as well as diabetes was also reported elsewhere (Aune et al., 2009; Martin et al., 2012).

In conclusion, the present small-scale cross-sectional study evidently unraveled the significant adverse effect of red meat on blood glucose level and hyperglycaemia/diabetes as well. There are a few limitations in the present study. Firstly, how meat consumption influences cardiometabolic risk, including the possible effects of red meat consumption on diabetes. Secondly, the study also did not consider the plausible association of genetic factors on hyperglycaemia and diabetes. Hence, further large-scale multidisciplinary approach considering genetic and other physiological factors in addition to lifestyle-related issues would be of importance among the population group for implementing future awareness and treatment programmes to gain success in the battle against this chronic disease.

Acknowledgements

The author is indebted to all the study participants for their wholehearted help during the data collection. The author is immensely thankful to Professor Barun Mukhopadhyay for his guidance in every step of the study. Indian Statistical Institute is gratefully acknowledged for financial support to conduct the study.

References

Alberti, KG, and Zimmet PZ 1998, 'Definition diagnosis and classification of diabetes mellitus and its complications. Part I: Diagnosis and classification of diabetes mellitus, provisional report of a WHO consultation'. *Diabetes Medicine,* Vol. 15, pp. 539-553.

American Diabetes Association 2010, 'Standards of medical care in diabetes – 2010'. *Diabetes Care,* Vol. 33 (Suppl) no.1, pp. S11-S61.

Aune, D, Ursin, G and Veieod, MB 2009, 'Meat consumption and the risk of type 2 diabetes: A systematic review and meta-analysis of cohort studies'. *Diabetologia,* Vol. 52, pp. 2277-2287.

Cho, NH, Shaw, JE, Karuranga, S, Huang, Y, da Rocha Fernandes, JD, Ohlrogge, AW and Malanda, B 2018, 'IDF diabetes atlas: Global estimates of diabetes prevalence for 2017 and projections for 2045'. *Diabetes Research and Clinical Practice,* Vol. 138, pp. 271-281.

Dowse, GK, Zimmet, PZ, Gareeboo, H, George, K, Alberti, MM and Tuomilehto, J 1991, 'Abdominal obesity and physical inactivity as risk factors for NIDDM and impaired glucose tolerance in Indian, Creole and Chinese Mauritians'. *Diabetes Care,* Vol. 14, pp. 271-282.

Harvard Magazine 2018, https://www.harvardmagazine.com/2012/01/a-diabetes-link-to-meat published in 2012. Accessed on 05.05.2018.

International Diabetes Federation 2006, *Diabetes atlas.* 3rd edition. Brussels, International Diabetes Federation.

Kaprio, J, Tuomilehto, J, Koskenvuo, M, Romanov, K. Reunanen, A and Eriksson, J 1992, 'Concordance for type 1 (insulin-dependent) and type 2 (non-insulin-dependent) diabetes mellitus in a population-based cohort of twins in Finland'. *Diabetologia,* Vol. 35, pp. 1060-1067.

Kurotani K, Nanri A, Goto, A, Mizoue, T, Noda M, Oba S, Kato M, Matsushita Y, Inoue M, Tsugane S; Japan public health center-based prospective study group. 2013, 'Red meat consumption is associated with the risk of type 2 diabetes in men but not in women: a Japan public health center-based prospective study'. *British Journal of Nutrition,* Vol. 110, pp. 1910-1918.

Levenstein, S, Prantera, C, Varvo, V, Scribano, ML, Berto, E, Luzi, C and Andreoli, A 1993, 'Development of the perceived stress questionnaire: A new tool for psychosomatic research'. *Journal of Psychosomatic Research,* Vol. 37, pp. 19-32.

Lohman, TG, Roche, AF, and Martorell, R 1988, *Anthropometric standardization reference manual.* Champaign, Illinois: Human Kinetics Books.

Martin, L, Tondeur, L, Fagherazzi, G, de Lauzon-Guillain, B, Boutron-Ruaualt, MC, and Clavel-Chapelon, F 2012, 'Processed and unprocessed red meat consumption and incident type 2 diabetes among French women'. *Diabetes Care,* Vol. 35, pp. 128-130.

Micha, R, Wallace, SK and Mozaffarian, D 2010, 'Red and processed meat consumption and risk of incident coronary health disease, stroke, and diabetes: A systematic review and meta-analysis'. *Circulation,* Vol. 121, pp. 2271-2283.

Mohan, VS, Sandeep, R, Shah DB and Varghese, C 2007, 'Epidemiology of type 2 diabetes: Indian scenario'. *Indian Journal of Medical Research,* Vol. 125, pp. 217-230.

Risley, Herbert Hope 1928, (Reprinted 1999), *The Gazetteer of Sikkim*. Delhi, Low Price Publications.

Sacks, DB, Arnold, M, Bakris, GL, Bruns, DE, Horvath, AR and Kirkman, MS 2011, 'Guidelines and recommendations for laboratory analysis in the diagnosis and management of diabetes mellitus'. *Diabetes Care,* Vol. 34, pp. 61-99.

Sarkar, S and Mukhopadhyay, B 2008, 'Perceived psychosocial stress and cardiovascular risk: Observations among the Bhutias of Sikkim, India'. *Stress and Health,* Vol. 24, pp. 23-34.

Steinbrecher A, Erber E, Grandinetti A, Kolonel LN, Maskarinec G. 2011, Meat consumption and risk of type 2 diabetes: the multiethnic cohort. *Public Health Nutrition,* Vol. 14, pp. 568-574.

Tuomilehto, J, Rastenyte, D, Qiao, Q and Jakovljevic, D 2004, Epidemiology of macrovascular disease and hypertension in diabetes mellitus. In: R. A. De Fronso, E. Ferrannini, H. Keen and Paul Z. Zimmet, [Eds.] *International Textbook of Diabetes Mellitus*, 3rd edn. Milan, John Wiley, and Sons, pp. 1345–1370.

World Health Organization 2014, *Global health estimates: Deaths by cause, age, sex and country, 2000-2012.* Geneva. WHO.

Zimmet, PZ 2001, 'Global and societal implications of the diabetes epidemic'. *Nature,* Vol. 414, pp. 782-787.

Index